Landmarks in Developmental Biology 1883–1924

Springer
Berlin
Heidelberg
New York
Barcelona
Budapest
Hong Kong
London
Milan
Paris
Santa Clara
Singapur
Tokyo

Klaus Sander

With Contributions from

S. J. Counce
P. E. Fässler
J.-L. Fischer
K. Nübler-Jung and D. Arendt

Landmarks in Developmental Biology 1883–1924

Historical Essays from Roux's Archives

Springer

Professor Dr. rer. nat. Klaus Sander
Universität Freiburg
Institut für Biologie I (Zoologie)
Albertstr. 21 a
D-79104 Freiburg
Germany

ISBN-13:978-3-642-64428-3

Library of Congress Cataloging-in-Publication Data.
Sander, Klaus. Landmarks in developmental biology, 1883–1924 : historical essays from Roux's archives / Klaus Sander ; with contributions from S. J. Counce ... [et al.]. p. cm.
Includes bibliographical references (p.).
ISBN-13:978-3-642-64428-3 e-ISBN-13:978-3-642-60492-8
DOI: 10.1007/978-3-642-60492-8

1. Developmental biology. I. Roux's archives of developmental biology. II. Title.
QH491.S26 1997 574.3–dc20 96-35278 CIP

Softcover reprint of the hardcover 1st edition 1997

Cover design: de'blik, Berlin
Typesetting: Camera ready from the journal Roux's Archives of Developmental Biology
SPIN: 10483111 27/3020 – 5 4 3 2 1 0 – Printed on acid-free paper

Preface

The essays in this collection were published in *Roux's Archives of Developmental Biology* from 1991 – to celebrate Volume 200 – to 1996, a century after the journal was founded by Wilhelm Roux as the *Archiv für Entwickelungsmechanik*. They are not intended to provide a comprehensive history of the early decades of developmental biology, nor are they for the historical expert who would certainly criticize the lack of references (and of due reverence) to the secondary literature. Rather, the essays were to serve as titbits, concise and centered around a single problem or paper, hopefully tempting developmental biologists of all ranks to read them, despite tight time schedules that usually discourage all but reading the most recent literature (and hopefully reflecting on it).

The stimulus inducing these essays came from some of my co-editors, who suggested that *Roux's Archives* should occasionally publish papers from the pioneering days of developmental biology. The interest in this period has been increasing for some time, apparently because several of the old (i.e. timeless) topics can now be approached with new, especially molecular, methods. However, many obstacles threatened to marr this project; for instance, the sheer bulk of nearly all papers under consideration. The final and negative decision, however, came from the knowledge gained during many years of teaching in the field – namely, that the modern reader would soon become frustrated because of terminological and conceptual discrepancies between then and now; discrepancies that could be resolved only with some background knowledge on the paper's context. Therefore, each translated paper would have required an introductory essay. In the end, then, the essays themselves seemed a preferable substitute. These are based almost exclusively on the detailed analysis of both the original papers and the conceptual environment in which they were written – with little consultation of secondary sources but first-hand knowledge of modern developmental research serving for comparison.

Most of the 25 contributions to this volume encompass only three pages each, in keeping with the "titbit" strategy. Yet three – declared *Centennial Essays* with respect to the occasion – are of wider scope and greater volume. They comprise a review of the papers published in the *Archiv für Entwickelungsmechanik* during the initial decades of developmental biology, mirroring contemporaneous foci of interest; a history of the stereo microscope, arguably the most important prerequisite of experimental research on development (and exactly as old as the *Archiv*); and, somewhat straining the time period, a review of pre-molecular speculations on a phylogenetic axis reversal between proto- and deuterostomian embryos, appearing appropriate in these days when molecular developmental biology has discovered evolution – with amazing insights resulting.

The authors figuring in these essays were mostly German speakers. This is not due to wilful selection favouring the birthplace of *Roux's Archives* – rather, it reflects the fact that the journal was conceived, and came to thrive, in Germany, because interest in the causes of ontogenesis was then stronger in that country than anywhere else. The resulting predominance of German research in nascent developmental biology is well reflected by the reference lists that E.B. Wilson selected 1896 for his seminal *Cell in Development and Inheritance*, as I have documented in a recent review (Int J Dev Biol 40:7–20).

Last but not least I would like to express my gratitude to the staff of Springer-Verlag and particularly those involved with Roux's Archives, who over the years were helpful in many ways, and to the friends and colleagues who generously have contributed some essays of their own: Sheila J. Counce, Durham/NC, Jean-Louis Fischer, Paris, Katharina Nübler-Jung and Detlev Arendt as well as Peter E. Fässler, Freiburg i.Br. May our joint efforts help students and adepts of developmental biology to understand and appreciate the achievements of a generation that has fundamentally shaped our fascinating discipline.

Freiburg i.Br., November 1996 Klaus Sander

Contents

List of Contributors

S. J. Counce

Department of Cell Biology, Duke University Medical Center,
Durham, NC 27710, USA

P. E. Fässler

Albert-Ludwigs-Universität Freiburg, Institut für Biologie I (Zoologie),
Albertstr. 21 a, D-79104 Freiburg, Germany

J.-L. Fischer

Centre Alexander Koyré, Muséum National d'Histoire Naturelle,
Pavillon Chevreul, 57 rue Cuvier, F-75231 Paris Cedex 05, France

K. Nübler-Jung, D. Arendt

Albert-Ludwigs-Universität Freiburg, Institut für Biologie I (Zoologie),
Albertstr. 21 a, D-79104 Freiburg, Germany

K. Sander

Albert-Ludwigs-Universität Freiburg, Institut für Biologie I (Zoologie),
Albertstr. 21 a, D-79104 Freiburg, Germany

ESSAY

Wilhelm Roux and his programme for developmental biology

Klaus Sander

Roux's *Archives of Developmental Biology*, founded nearly 100 years ago as *Archiv für Entwickelungsmechanik der Organismen*, was the first journal devoted exclusively to the causal analysis of ontogenetic development. After a varied history (see back of volume title page), it is now entering Volume 200. On this occasion, the editor and the publisher have decided to initiate a series of essays highlighting problems and achievements in developmental biology in the discipline's early years. We are encouraged in this venture by the increasing interest taken by practising developmental biologists in various countries in the history of their discipline and in the views of its pioneers – views that in our day may be appreciated in a more detached and (hopefully) more objective vein than by earlier generations of developmental biologists, who had to struggle with the pioneers' inevitable errors while taking their achievements for granted.

We also hope that, by devoting subsequent essays to individual key papers, we might enable readers not conversant with "continental" languages and, especially, with nineteenth century "Academic German" to form an opinion on what those papers are really about. Our plans were triggered by a suggestion that we should publish complete translations of landmark papers that are so far available in English to only a very limited degree (e.g. Willier and Oppenheimer 1964). This, however, would be impractical because of the sheer volume of most of the papers in question and also because of terms and arguments in them that cannot be understood or appreciated without a certain background knowledge of nineteenth century biology.

Had we decided on the conventional basis of marking decades and centuries (see Bynum and Heilbron 1988), Hans Driesch and his 1891 paper on experimental twinning in the sea urchin would have been prime candidates for the first essays. However, we decided in favour of Wilhelm Roux – not so much because he was the founder of this journal but rather because he conceived and explicitly set forth the aims and principles of a new discipline, called *Entwickelungsmechanik*, that was to grow into developmental biology. "It is safe to say that all the analytical embryology of the late nineteenth and of the twentieth century has built upon foundations laid by Roux" (Willier and Oppenheimer 1964).

Wilhelm Roux (1850–1924) was born in the small university town of Jena, about the same time as a young mechanic named Carl Zeiss set up his workshop there. The family's French name was carried to Germany in 1687 by a lawyer who, being a Huguenot, had to flee from Grenoble because of religious persecution. Roux's father earned his living by teaching Jena students the then fashionable art of fencing. His strong influence may explain some of his son's stern rigidity in defending his scientific views and his (sometimes unwarranted) claims to priority which made him few friends but lots of enemies. "That Roux had a substantial impact upon his peers and upon contemporary embryology there can be little doubt, but it is equally clear that he was a ferocious propagandist for his own accomplishments" (Churchill 1975).

After completing the medical studies for which he had qualified himself by a course of home study (which also bears witness to his tenacity and determination), he embarked on a career in medical anatomy, picking up some philosophy, physics and chemistry along the way. In 1886 he became associate professor of anatomy at Breslau (now Wroclaw), and within a few years was master of the "Institut für Entwickelungsgeschichte und Entwickelungsmechanik" established for him by a far-sighted Prussian administrator. Although this made him the first and the only professor of *Entwicklungsmechanik* in the world, he left Breslau after less than 2 years in order to head the new institute of anatomy in the Tyrolean capital, Innsbruck. In 1895 he took up the corresponding position in Halle in central Germany, where he spent the remainder of his life.

Reading Roux requires a mind trained in the writings of (mainly German) philosophers, perseverance in analysing voluminous papers, and determination to cope with a terminology that puts some strain even on readers well-versed in classical languages. For instance, who would guess the meaning of "hemioo-holoplast"? It is Roux's term, which he first used in 1892, for a complete embryo derived from half an egg, i.e. from a single blastomere of the two-cell stage. It was his leaning to philosophy and terminology that made Roux coin the term "*Entwickelungsmechanik*", which has been misinterpreted as frequently as his half-embryos (of which more later). "*Mechanik*" here does not refer to the branch of physics concerned with movements and forces; rather – as used by eighteenth century philosophers, foremost Immanuel Kant – it stands for "natural causation". The alternative designation "developmental physiology", which was favoured for example by Driesch in his later years, was rejected by Roux with the (then as now) irrefutable argument that mainstream physiology is interested in the functioning rather than the generation of organismic structures.

Roux's predilection for philosophical terminology is rooted deeply in his approach to science. With him, the postulate that any experiment should be preceded by intense reflections about the questions it was to answer, and indeed about the possible answers, was not lip service but derived from conviction: "... I invested years of pondering in order to track down the possibilities of how a structure as complex and typically shaped as the chick could arise from a relatively or seemingly simple state without external formative influences" (Roux 1885, p. 419; all translations by the present writer). Indeed, Roux published his reflections on what *Entwicklungsmechanik* should be and do (Roux 1885) before communicating any of his embryological experiments, except for a paper on the effects of gravity. Hans Driesch – who ought to know – considered Roux a philosopher by character rather than a natural scientist. However, Roux's philosophy was not Driesch's vitalism for he held that "in the material course of embryonic development nothing metaphysical must be considered", and that one must be convinced "that these events are throughout subject to the law of causality" (Roux 1885, p. 419). General philosophical reflections can merely serve "as heuristic principles for our tedious exact special research" (Roux 1885, p. 420); the method for this research is the analytical experiment, "and in order to perform this, it has to be preceded by analytical thinking" (Roux 1889, p. 9).

Roux was by no means the first experimental embryologist. During his prime years several notable scientists, including L. Chabry, O. Hertwig and G. Born, started experimenting on embryos with the aim of understanding their normal development rather than teratogenesis, which had been the concern of previous generations (see Oppenheimer 1968). However, none of these contemporaries wrote as explicitly and repeatedly about their general aims in addition to their specific problems, as did Roux. "He had dramatized experimentation on the embryo in a way that his peers had not, and in a unique manner he had made explicit that a true experiment was not just a manipulation of the eggs but included a mental dissection of development into a number of hypothesized factors. In this respect his analysis of development into dependent and independent factors promoted a causal understanding" (Churchill 1975).

Let us illustrate this evaluation with a few quotations from Roux's early programmatic writings (Roux 1885, 1889). Roux explains the need for his new discipline by pointing out that the existing branches of anatomy – descriptive, physiological(!), ontogenetic and comparative – are incomplete: "... we are lacking a large part, not to say the best part, of knowledge required for complete insight; we are lacking the knowledge of the proximate causes of the development of these structures ... We do not know which forces are localized in the fertilized egg and in which arrangement, so as to be able to initiate the development of the individual. We do not know which combinations of forces effect the further course of development; in short, we do not know why the simply structured egg gives rise to a highly complex and typically constructed organism, and why the organism that forms in this way remains rather unaltered over quite some time in spite of the continuous metabolism of its constituents" (Roux 1889, p. 5). In order to provide this knowledge, *Entwicklungsmechanik* has to proceed along two pathways, namely description and experimental analysis. "We shall never see either approach completed; however, we shall have to attend continuously to both in order to approach our aim in both ways; the required combination of both tasks might be named *Entwickelungsmechanik des Embryo*" (Roux 1885, p. 413). Roux then ponders whether at the present level of ignorance about the germ plasm and even about the relevant properties of organized matter in general any meaningful experiments can be made. Resorting to a black-box approach he arrives at a positive answer. Pointing out that in the lowest (unicellular) organisms there is cell division but no subsequent development, he concludes that ontogenetic "development is not a necessary consequence of life per se, but rather something added to the secret basis of life, something which therefore perhaps can be analysed successfully for a certain stretch of the way without having gained previous insight into the basis itself" (Roux 1885, p. 421).

The experiments to be performed must of course be directed by analytical thinking, which has indeed thrown up many questions of causality that can be approached with the experimental methods available. Almost all these questions "lead, when followed up, to one and the same basic question, to an alternative which directs the causal interpretation of almost all developmental events into two essentially different tracks", namely independent and dependent differentiation (the latter named "correlative differentiation" at first).

It is under this alternative that Roux sets out to discuss a concept originating from another author, and he does so in a non-pugnacious style – a rare achievement for him. The author is Wilhelm His (1831–1904), the concept that of "organ-forming primordial regions" hailing from 1875 (see, for example, Moore 1987). This concept, contrary to what many text-books claim, was merely that of a fate map, and not that of a mosaic of determinants or determined regions. Roux discusses both alternatives with reference to the distinction made by himself. If the parts develop independently, that is, if development merely represents "metamorphosis" of invisibly preformed spatial complexity into visible shape, then the principle of W. His "obtains besides its descriptive character a simple causal meaning which indeed may be extended down to the fertilized or even the unfertilized egg cell ... If, however, development occurs essentially by interaction between many or all parts, the fertilized egg needs to consist of only a few different parts, which by mutual interactions gradually generate a great complexity. Development in this case essentially is *production of complexity, epigenesis* in our sense. There is a mutual cooperation of parts to form a whole, whereby the whole can then exert a regulating influence on the parts ... His's principle of organ-forming primordial regions, however, then has causal importance only in so far as it marks the locations of the resultants of manifold interactions, and it is of little value to project these locations onto the undivided or unfertilized egg and into the period of time before these interactions have taken place" (Roux 1885, p. 427ff).

Note that the latter alternative embodies as clear a description of "regulative" development as could be wished for. Moreover, it is this alternative that particularly invites experimentation, for "in establishing these correlations we are provided with a rich field of research that can be attacked with the technical means of our time" (Roux 1885, p. 428).

In consequence of these considerations, Roux in 1885 proposed three topics for causal research: "The mechanics of de novo production of complexity, the mechanics of the metamorphosis of deduced (i.e. invisible) complexity into perceptive complexity, and the mechanics of combining both these primary types ... since the processes which represent our special tasks are composed by the cooperation of all three" (Roux 1885, p. 419).

Over the years, Roux advertised his programme in many publications and with gradual changes or modifications, but in principle his message is contained in these early papers. His other means of promoting *Entwickelungsmechanik* was the founding of periodicals devoted to this disipline, among them the *Archiv für Entwickelungsmechanik der Organismen.* In the year of his retirement, the journal incorporated the name of its founder into the title, from which it was dropped only temporarily during a merger with another journal.

The editor and the publisher hope that they will be able to carry on under Roux's name even in these days of seemingly logarithmic multiplication of developmental journals. For Wilhelm Roux excelled in developmental biology "at a time when it made dramatic strides; he made substantial contributions to theories of development and inheritance; he proselytized for and philosophized about it; and he produced the journals and terminology for its dissemination. To the degree that he performed these tasks more zealously than his contemporaries, he was the titular sire of modern experimental embryology" (Churchill 1975).

References

Bynum WF, Heilbron JI (1988) Eighteen eighty eight and all that. Nature 331:27–30

Churchill FB (1975) Roux, Wilhelm. In: Gillispie CC (ed) Dictionary of scientific biography, vol 11, Scribner, New York, pp 570–575

Moore JA (1987) Science as a way of knowing – developmental biology. Am Zool 27:415–573

Oppenheimer JM (1968) Some historical relationships between teratology and experimental embryology. Bull Hist Med 42:145–159

Roux W (1885) Beiträge zur Entwickelungsmechanik des Embryo, Nr. 1 (Einleitung). Z Biol 21:411–524

Roux W (1889) Die Entwickelungsmechanik der Organismen, eine anatomische Wissenschaft der Zukunft. Festrede. Wiener medizinische Presse. Urban und Schwarzenberg, Vienna 1890. (Reprinted in Roux W, Gesammelte Abhandlungen, vol 1, 1895)

Willier BJ, Oppenheimer JM (eds) (1964) Foundations of experimental embryology. Prentice-Hall, Englewood Cliffs, N.J.

WILHELM ROUX
JENA 1850 · HALLE·S·1924

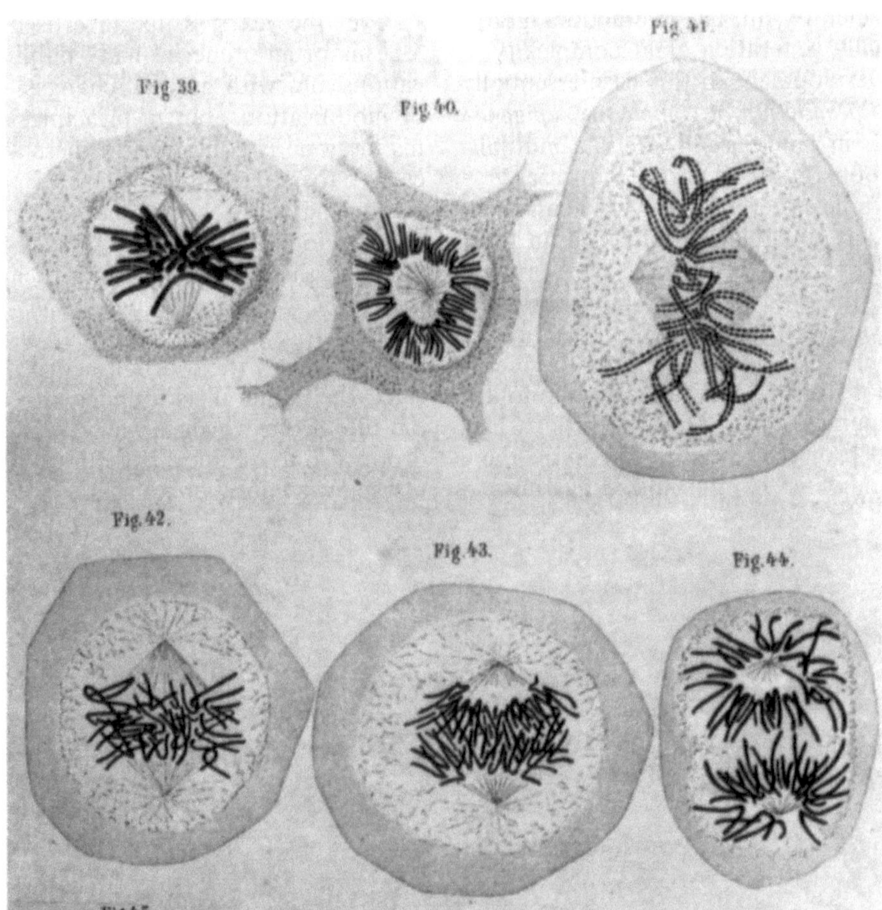
Fig. 39.
Fig. 40.
Fig. 41.
Fig. 42.
Fig. 43.
Fig. 44.
Fig. 45.

point of view, Flemming (a pathologist by profession) failed to comment on this all-important splitting of chromatids; all he discussed in 1879 was the possibility of differences between nuclear divisions in normal and in pathological situations. Moreover, the maxims he expressed in 1882 (p. VI) were stern: "He who in a new book first looks for new catchwords will miss here ... any hypotheses or theoretical views on cell division. All constructions which hitherto were made in this direction resemble groping in a dark room where nothing has been felt as yet with certainty. One will have to proceed with this; but here, where I attempt to present the facts as of to-day, I wanted to stay with the palpable."

Not so Wilhelm Roux. After briefly summarizing the complicated events described by Flemming and others, he stated his aim as follows: "We now wish to learn what this extensive play of changing shapes is good for, what is its use for the ultimate purpose of dividing the simple nucleus in two halves" (p. 6). After pointing out that the effort involved in indirect division (mitosis) must provide some advantage not to be gained by the simpler means of direct division, he continues: "If the purpose of nuclear division were just a simple halving of the mass of the nucleus and the spatial separation of the two halves, it would be evident that the course of indirect division represented an enormous detour for reaching a nearby goal, that is, it would be quite inappropriate. The verdict will be different if the aim of nuclear division is not a mere random halving of the nuclear mass, but a defined segregation of the qualities which together constitute this mass" (p. 6).
This conceals a terminological trap for the unwary: with Roux, "segregation of qualities" does not mean segregation of differing nuclear qualities into **different** daughter cells, as our (and Wilson's) usage of "qualitative division" implies; rather, segregation of qualities denotes their **equal** distribution to both nuclei! On the following pages, Roux clarifies this point beyond any reasonable doubt, by first going through a chain of finalistic arguments and then showing that the observed events are able to serve the inferred purpose, namely a "qualitative halving" of the chromatin. "Consider that a mixture of different substances in the nucleus should be halved in such a way that each half comprises one half of each of those substances" (p. 6). Roux's solution involves splitting up the mass into increasingly smaller fragments – in accordance with Flemming's erroneous belief that the chromosomes derive from a single continuous chromatin thread in the parent nucleus: "If this mechanical dissection continues until there are as many or even more fragments than there are qualities in the whole mass, and if we then bisect each fragment and put one half to one side and the other half to the other, then the sum of all fragments on one side will rather completely equal that on the other side, not only by its mass but also in its qualitative and percentile constitution" (p. 7). After a lengthy discussion on mechanisms that could best achieve the moving apart of fragment halves – including the idea that lining the fragments or granules up in a (chromosomal) thread (Fig. 1) would facilitate the task – Roux concludes: "If this is the correct meaning of mass division, then observation should prove that normally the daughter threads of any mother thread should never arrive on the same side but always be distributed to both sides, because otherwise that which in our opinion is the aim of the molecular division [halving of the granules and threads] would be abolished, and consequently this division would be superfluous" (p. 17). This postulate was verified by contemporaneous observations of Eduard Strasburger, whose student Emil Heuser published a corresponding diagram in the following year (Fig. 2, below). In the light of these quotations, one can but marvel at such statements as this one from Churchill (1975): "Obviously, Roux rationalized, such a complex and indirect division of the nucleus must serve some functional purpose; here perhaps was a mechanism for dividing the nucleus into two qualitatively **unequal** halves and for segregating these differences to the daughter blastomeres."

This brings us to Wilson's rendering of Roux's treatise, and to development. In writing his paper, Roux was evidently not bent on explaining development but on understanding the meaning of mitosis. Wilson (1896) renders this aspect correctly on p. 183, but on

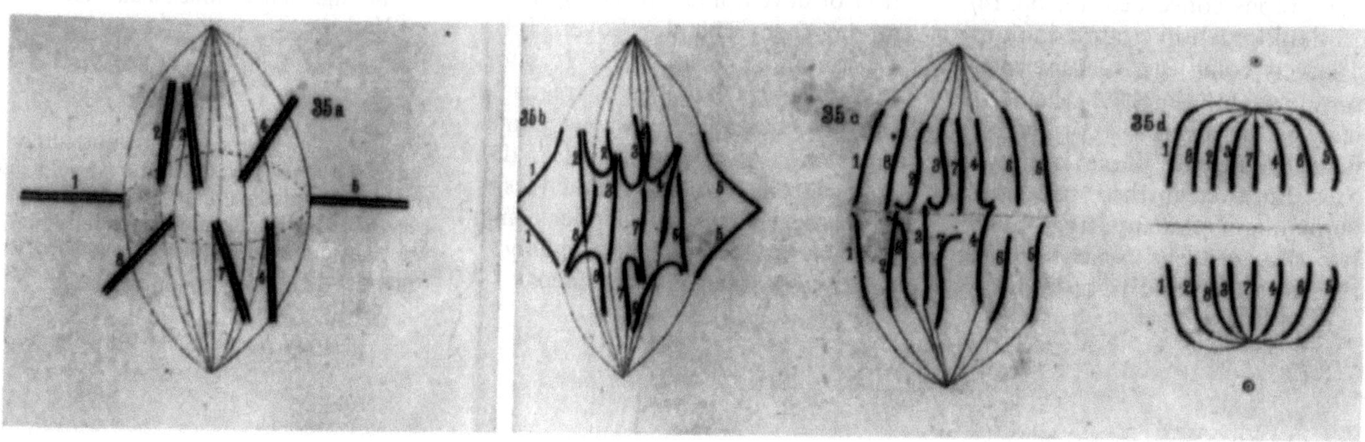

p. 304, when he turns to development, he alters both the tenor and the terminology of Roux's treatise: 'Roux assumes, as a fundamental postulate, that division of the [chromatin] granules may be either *quantitative* or *qualitative*. In the first mode the group of qualities represented in the mother granule is first doubled and then split into equivalent daughter groups, the daughter-cells therefore receiving the same qualities and remaining of the same nature. In "qualitative division", on the other hand, the mother group of qualities is split into dissimilar groups, which, passing into the respective daughter nuclei, lead to a corresponding differentiation in the daughter cells.' While this is compatible with the original – apart from the switch in terminology – the next sentence is not: "By qualitative divisions, occurring in a fixed and predetermined order, the idioplasm is thus split up during ontogeny into its constituent qualities, which are, as it were, sifted apart and distributed to the various nuclei of the embryo." This last sentence doubtless renders ideas of August Weismann (more in another essay); certainly, statements on the predetermined sifting of constituent qualities cannot be found in Roux's treatise of 1883 (to which Wilson refers his readers).

Actually, Roux devoted less than a single page in his 19-page treatise to "qualitatively unequal divisions", merely pointing out that the mechanism of indirect division might also support the unequal distribution of nuclear qualities. The paragraphs concerned (on pp. 14/15) look much like late additions to a largely completed manuscript and were evidently inserted when Roux had just discovered a correlation between cleavage planes and body axes. But even in these paragraphs he maintains that the frog's first two cleavage cells, which will give rise to the left and right body halves, respectively, ought to contain equal nuclear qualities, while the second cleavage divisions, separating the prospective anterior and posterior body parts, will probably be "qualitatively unequal."

During the four decennia following 1883, Roux mentioned nuclear division on innumerable occasions. Depending on the context, he stressed either the "qualitative halving" or "qualitatively unequal divisons." In his autobiography (Roux 1923), which has to be taken with more than a grain of salt, he ascribes the typical development of the body to unequal mitoses and dismisses as improbable the idea that all cells receive equal nuclei. On other occasions (e.g. Roux 1904, p. 124ff), he relegates unequal division to the footnotes while discussing equal division at length. In neither case did he fail to impress on the reader that it was Wilhelm Roux who had conceived the idea in question, which makes one suspect that the "ferocious propagandist" in him decided on each occasion which stance to take.

Be that as it may, neither view of nuclear division was historically as productive as Roux's indirect but compelling arguments that chromatin must embody a highly complex "metastructure", arguments that he deduced from his interpretation of mitosis as a mechanism for **equal** nuclear division. This claim, the first ever, of complexity hidden in the seemingly homogeneous chromatin profoundly influenced the thinking of August Weismann and of the next generation of developmental biologists, among them Theodor Boveri, E.B. Wilson and Hans Driesch. It is for this influence that Roux's treatise ought to be called classic. Its fame regarding "qualitative" nuclear divisions (sensu Wilson) is much less well deserved, resting as it does on a casual remark amplified by highly selective renderings of the actual content by later writers, foremost among them E. B. Wilson. Yet it was this fame that conferred on Roux's treatise the status of a "classic" – a status that is confirmed both by its lasting influence on textbook authors and, unless my suspicions are mistaken, by its being quoted without having been read.

References

Churchill FB (1975) Roux, Wilhelm. In: Gillispie CC (ed) Dictionary of scientific biography, vol 11. Scribner, New York, pp 570–575

Cremer T (1985) Von der Zellenlehre zur Chromosomentheorie. Springer, Berlin Heidelberg New York

Flemming W (1879) Über das Verhalten des Kerns bei der Zelltheilung. Virchows Arch [A] 77:1–29

Flemming W (1882) Zellsubstanz, Kern und Zelltheilung. Vogel, Leipzig

Heuser E (1884) Beobachtungen über Zellkerntheilung. Bot Zentralbl 17:27ff [several sections]

Oppenheimer J (1965) Questions posed by classical descriptive and experimental embryology. In: Moore JA (ed) Ideas in modern biology. Natural History Press, Garden City/NY, pp 205–207

Roux W (1883) Über die Bedeutung der Kerntheilungsfiguren – Eine hypothetische Erörterung. Engelmann, Leipzig

Roux W (1904) Die Entwicklungsmechanik, ein neuer Zweig der biologischen Wissenschaft. Engelmann, Leipzig

Roux W (1923) Wilhelm Roux in Halle/Saale. In: Die Medizin der Gegenwart in Selbstdarstellungen 1:141–206. Meiner, Leipzig

Sander K (1991) Landmarks in developmental biology: Wilhelm Roux and his programme for developmental biology. Roux's Arch Dev Biol 200:1–3

Sturtevant AH (1966) A history of genetics. Harper and Row, New York

Wilson EB (1896) The cell in development and inheritance. Macmillan, New York

All translations are by the present author. Italics are the respective author's; boldface type and words in square brackets have been added for better understanding. Figure 1 is from Flemming (1882), and Fig. 2 from Heuser (1884).

ESSAY

Landmarks
in Developmental Biology (3)

Wilhelm Roux on embryonic axes, sperm entry and the grey crescent

Klaus Sander

In the very year, 1883, when Wilhelm Roux was pondering on the meaning of mitosis (see Sander 1991), he started experimenting on amphibian eggs. The first problem he set out to solve – or rather, the first question he discussed in writing – was the orientation of the prospective embryo in the egg. In amphibians this is a formidable problem, partly owing to the highly complex gastrulation movements, which in some respects have remained enigmatic to this day. Indeed, on trying to ascertain dorsoventral and antero-posterior polarity in the frog's egg, Roux was confused by aberrant gastrulation movements in quite a spectacular way. Irrespective of this, he gained remarkable and still valid insights concerning the prospective median plane in frog eggs and the epigenetic factors that bring about its orientation – the latter study (Roux 1887) being adorned with further speculations on equal and unequal mitoses during early cleavage.

The first paper was "On the time of determination of the main directions in the frog embryo" (Roux 1883a). Roux justifies his topic as follows: "Whether the directions of the various formal and qualitative processes of differentiation that we know of become determined together with these processes themselves, or whether they receive their normation much earlier by other occurrences, it will always be of interest for us to know at what time the direction in which each process occurs is laid down, because only then can we search, with a chance of success, for the true causes of this normation" (Roux 1883a, p. 3). However, "since in the frog's egg, as in all telolecital eggs ... the position of the yolk always determines the belly side, only the determination of the median plane had to be studied. ... As the blastula does not reveal any [circumferential] orientations, I went back directly to cleavage and actually to its beginnings. I thus asked: Is there any constant relation between the orientation of the first cleavage plane and of the embryo's median plane, or are they independent of each other?" (Roux 1883a, p. 9–10).

To answer this question, Roux chose the edible green frog (*Rana esculenta*) whose cleavage, at least in early spawnings of the strain he used, follows a bilaterally symmetrical pattern to begin with (see Figure below). Immobilizing by various means the egg cell and its jelly envelope, and marking on the substrate the orientation of the first cleavage furrow (labelled I), he soon found that its plane coincided very reliably with the embryo's median plane, as revealed by the dorsal groove at neurulation.

This fact established, Roux posed the question of head versus tail. The subsequent location of the respective body poles, he found, could also be predicted from the early cleavage pattern, because the second cleavage furrow (II) arises somewhat off-centre, yielding two larger and two smaller blastomeres. The latter, as Roux found out by observing his immobilized eggs, mark the position that is later occupied by the tail end. Scrutinizing early cleavage stages, he made an additional discovery, which proved invaluable for later work on amphibians, and particularly on those species and strains that lack orderly cleavage patterns.

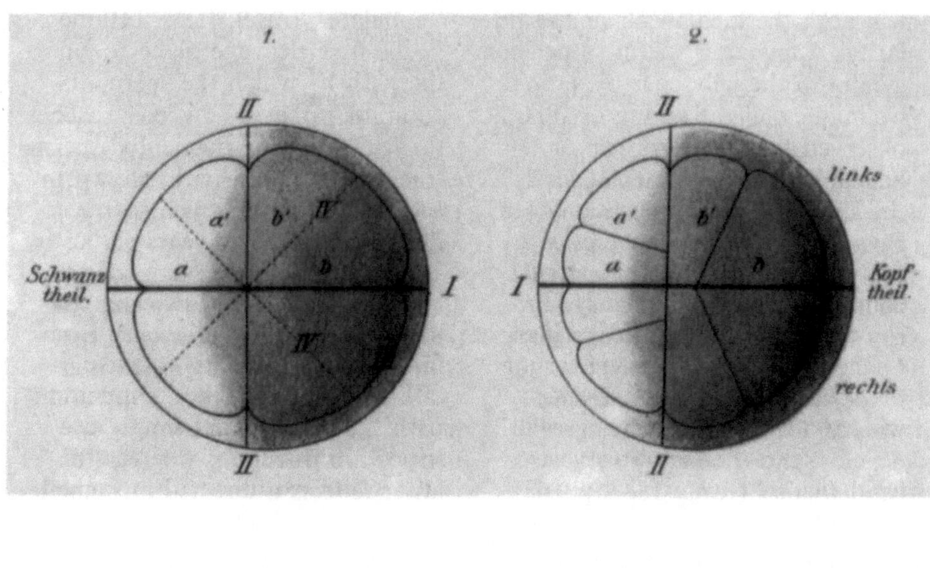

He writes: "The excentric position of the second cleavage plane is constantly determined ... beforehand. Owing to the internal differences in the specific weight of its materials, in every case the egg orients in such a way that the center of the brown pole [center of the pigmented cap] is not exactly at the top but inclined somewhat to one side; on the opposite side, therefore, a small part of the marginal zone of the white pole [white hemisphere] makes its appearance. It is invariably towards this side that the second furrow is positioned excentrically" (Roux 1883a, p. 19). In a later reproduction (shown here) of his original line drawings, Roux added the excentric pigmentation exposing, at the prospective "Schwanztheil", an unpigmented crescent-shaped stretch of the white hemisphere. Some years later (Roux 1888, p. 117), he noticed that in *Rana fusca* "the dark hemisphere brightens in form of a crescent-shaped grey seam abutting the white hemisphere" – enter the grey crescent!

After knowing that under normal conditions the orientation of the median plane and anterior-posterior polarity seem to be determined before or at first cleavage, Roux turned to analysing the orienting factors (Roux 1887). Because the future median plane and axial polarity within this plane are predictable just after fertilization, when the brown cap has tilted, Roux guessed that the process of fertilization might determine the direction of tilting and thereby establish both the median plane and polarity. Applying a sperm suspension locally to the jelly coat by various techniques, in the spring of 1886 he observed that the dark cap or hemisphere always tilted towards the side of sperm application: "In eggs of *Rana fusca* and *esculenta* that are not subjected to external constraints, the first cleavage furrow, and with it the median plane of the embryo, are passing through the sperm entrance site selected at will, and that side of the egg will develop into the embryo's ventrocaudal side" (Roux 1887, p. 163)

This "thesis", as Roux called it, is interesting for two reasons in addition to its main message. First, Roux had meanwhile noticed that by constraining the uncleaved egg in an abnormally tilted position, one can overcome the sperm's orienting influence: "With strong tilting of the egg axis, that side towards which the dark ... pole is inclined will always form the ventrocaudal side of the embryo" (Roux 1887, p. 197). Second, apparently he had changed his mind about the axes in the median plane. While in 1883, as quoted above, the pigmented cap was considered dorsal and tilted towards the future anterior end ("Kopftheil" in the Figure), in 1887 Roux assigned it a prospective ventro-caudal fate. The reasons for this change of mind will be explored in a later essay.

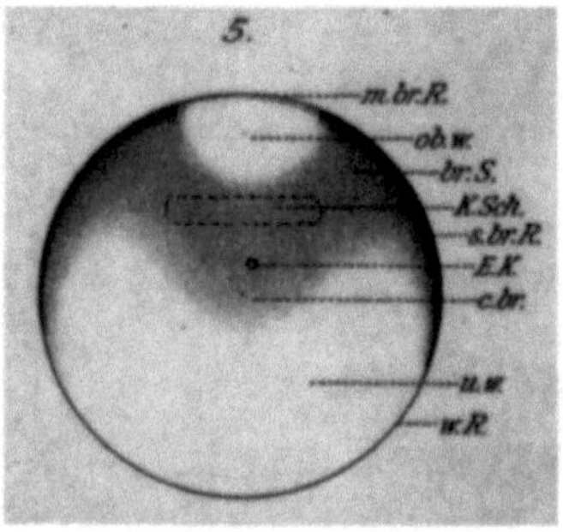

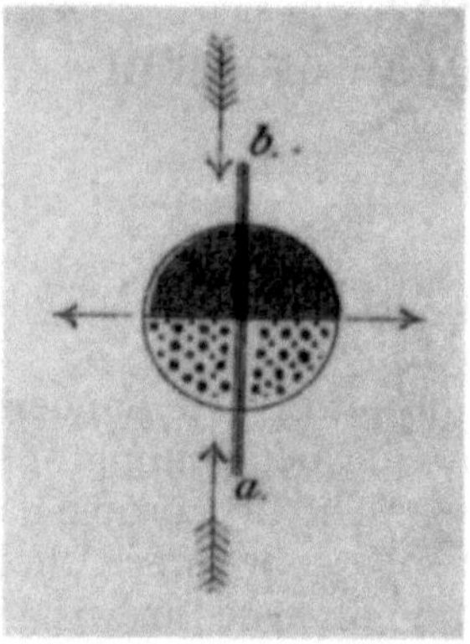

We now turn to the role that Roux ascribed to the sperm and its nucleus in establishing symmetry and polarity of the future body. Having noted constant spatial relations between sperm entry site and symmetrization, he also wanted to know the mechanisms that linked the two. He therefore killed eggs before and during first cleavage by heating them to 80° C (no higher, because they would have shrivelled up) and sectioned them on a freezing microtome or after hardening in alcohol. Sections of unfertilized eggs yielded the aspect shown in the figure marked 5 (for abbreviations see end of reference list). The dashed contour marked "K.Sch." outlines the region where the pronuclei meet and "copulate" (fuse). Eggs sectioned in the first cleavage plane reveal the path by which the spermatozoon, or rather its nucleus, reaches this site: it is the pigmented sperm track (penetration path) shown in Fig. 6. This path, which had been discovered some 10 years earlier by Van Bambeke in Belgium, consists according to Roux of two components, the radially oriented "penetration path" and its "nucleopetal" continuation, the "copulation path". The former is simply centripetal in direction, whereas the latter reflects a directed movement of the sperm nucleus towards the egg's pronucleus; the two components may form a conspicuous angle, as documented in Figs. 6 and 13. It is the course of the copulation path and the resulting "copulation axis" of the pronuclei (see below) that, according to Roux, determine the median plane. One of the reasons for his claim is evident from the sketch in Fig. 13 (showing a polar view at the two-cell stage), where only the copulation path runs parallel to the furrow marking the median plane – a rather weak argument even in Roux's time.

Actually, the copulation path may have acquired its dominant role in Roux's scheme through a set of speculations that lead us back to mitosis. If the sperm nucleus is heading for the egg pronucleus right in the future median plane, the contents of each copulating

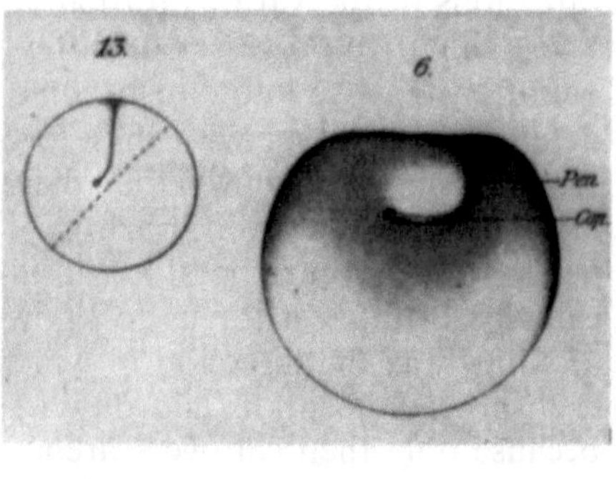

pronucleus might be viewed as forming a hemisphere bisected by that plane (marked a–b in the sketch below Fig. 5). A mitotic spindle arranged at right angles to this plane – as required for first cleavage to occur along the midline – would then separate equal portions of each pronucleus and distribute these portions to the daughter cells as indicated by the horizontal arrows. This mechanism, which Roux discusses somewhat less simplistically than I render it here for the sake of brevity, would help to ensure that both sides of the future body receive **equal** nuclear qualities.

Unequal nuclear division, envisioned for the second cleavage cycle in the treatise on mitosis (Roux 1883b, see Sander 1991), posed a special problem, for meanwhile Roux had observed some cases where the first instead of the second furrow appeared in an excentric position and was found to separate anterior from posterior blastomeres, while the second furrow cut through both poles and marked the median plane. Roux accounted for these cases ad hoc and with fairly scholastic arguments: "I think ... that here the type of [nuclear] segregation that should come second was released as the first: a very interesting example of ... prestabilized self-regulation and at the same time a sign of the superiority of the mechanical tendency for [unequal] segregation ... over the simultaneously present tendency for halving the nuclear material at first [division]" (Roux 1887, p. 199). The conceptual interest of these arguments rests not in their sophistry but in a detail evident mainly from the context, namely that the unequal segregation of nuclear qualities is triggered by the surrounding cytoplasm – a profound difference from Weismann's contemporaneous ideas on the autonomous segregation of determinants.

In closing, let us look at the status of knowledge before Roux's time and in our own day. Roux's claim that sperm entry can (and normally does) establish and polarize the median plane of the frog embryo has obviously stood the test of time, as has his discovery that the effects of sperm entrance can be overridden by gravity; a century later, analysis of the underlying mechanisms is evidently well on its way (see, for example, Elinson 1991). When we turn to the period before Roux, we find that George Newport in England had noticed the sperm's orienting influence some time before 1854. Roux convincingly claims to have learned of this predecessor only after completing his own investigations (Roux 1887, p. 210), but when he condescended to acknowledge Newport's priority it was "although his experiments were not sufficiently numerous and varied enough for obtaining entirely reliable results. These and some other important observations of Newport seem to have met with no understanding among his contemporaries and therefore were entirely lost from memory ... Thus it came about that they remained infertile for science and were re-discovered only when adequate research was inaugurated methodically, establishing the facts anew: regrettably the usual fate of solitary achievements that exceed contemporary imagination" (Roux 1887, p. 211). Sad but true – unrecognized discoveries solicit respect in later times but fail to influence the course of history. This is what Roux might have learned from Newport's vain achievements – had he, with his omnipresent craving for recognition, ever needed that lesson.

References

Elinson R (1991) Polarity of the surface and cortex of the amphibian egg from fertilization to first cleavage. J Electron Microsc Tech 17:369–383

Roux W (1883a) Über die Zeit der Bestimmung der Hauptrichtungen des Froschembryo – Eine biologische Untersuchung. W Engelmann, Leipzig

Roux W (1883b) Über die Bedeutung der Kerntheilungsfiguren – Eine hypothetische Erörterung. W Engelmann, Leipzig

Roux W (1887) Beiträge zur Entwickelungsmechanik des Embryo. 4: Die Richtungsbestimmung der Medianebene des Froschembryo durch die Kopulationsrichtung des Eikernes und des Spermakernes. Arch Mikrosk Anat 29:157–212

Roux W (1888) Beiträge zur Entwickelungsmechanik des Embryo. 5: Über die künstliche Hervorbringung halber Embryonen durch die Zerstörung einer der beiden ersten Furchungszellen, sowie über die Nachentwicklung (Postgeneration) der fehlenden Körperhälfte. Virchows Arch 114:113–291

Roux W (1895) Gesammelte Abhandlungen über Entwickelungsmechanik der Organismen, vol 2. W Engelmann, Leipzig

Sander K (1991) Wilhelm Roux's treatise on "qualitative" mitoses – a "classic" by either definition. Roux's Arch Dev Biol 200:61–63

All translations by the present author; words in square brackets have been added for better understanding. The first figure is from Roux 1883a as modified in Roux 1895, the other figures are from Roux 1887. Abbreviations: *m.br.R.*, central (upper) brown cortex; *ob.w.*, upper white yolk; *br.S.*, brown lateral yolk; *K.Sch.* nuclear fusion site in yolk (see text); *s.br.R.* lateral (lower) brown cortex; *E.K.*, egg nucleus [actually pronucleus]; *c.br.*, central brown yolk; *u.w.*, lower white yolk (coarsely grained); *w.R.*, white cortex; *I–IV* successive furrows.

ESSAY

Landmarks
in Developmental Biology (4)

When seeing is believing: Wilhelm Roux's misconceived fate map

Klaus Sander

"*Entwickelungsmechanik*, like every new approach in science, will have to acquire its deserved status step by step. Nonetheless, posterity might well be amazed on learning that the now ruling descriptive school ignored the secure findings of *Entwickelungsmechanik* for a long time, until descriptive research had led to the same views, and especially that this school placed more confidence in the latter kind of findings than in the former. This will bear permanent testimony to the insufficient appreciation that those researchers had for the value of experiments." This was Wilhelm Roux's verdict at the end of a discussion of gastrulation and the fate map in frogs (Roux 1892, p. 441). In the preceding paragraph he had summarized his conclusions on these topics, and at the same time paved the way for the verbal showdown just quoted: "The conclusion, based on three different types of experiment, that gastrulation in the frog occurs by way of bilateral epiboly and concrescence on the lower side of the egg and that the outer face of the blastoporal lip is the site of the neural folds, is therefore so secure that its certainty could neither be increased by the consent of descriptive researchers, which is per se pleasing, nor could it be diminished by their opposition; the facts thus established must rather be viewed as the firm foundations of our knowledge of developmental processes, meaning at the same time that all views clearly incompatible with these facts can for sure be called mistaken" (Roux 1892, p. 440/41).

Had he lived to see fate maps based on cell marking methods, Roux (had he been so disposed) might well have cursed himself for these lines, because never before or afterwards did he as dogmatically and emphatically endorse what in the end turned out to be a wrong option.

How, then, did Roux come to acquire his mistaken beliefs? In the course of his observations and experiments on frog eggs begun in 1882, he had occasionally observed eggs that failed to gastrulate properly. The same type of failure also occurred sometimes after he had damaged one blastomere at the four-cell stage (see the next essay). In either case, the material of the dark hemisphere stopped moving before it had completely overgrown the white cell mass derived from the vegetal region of the egg cell. The blocked rim of the dark cap had then turned into half a neural plate on either side of the persisting white yolk plug. With time, these halves could slowly approach each other, so that sometimes they would remain apart only in a small region of the trunk (Roux 1888a, b). As Roux provided his readers merely with the sketch of a single, quite severe, case (marked Fig. 4 in the illustration below), I have added on the next page a series of embryos showing milder degrees of the same "split axis" malformation. These figures were published by Roux's arch enemy Oskar Hertwig in 1892, as were the transverse sections reproduced below them. The upper section (3) comes from the embryo above it (marked 6), passing through its midbody region. It shows that on either side of the yolk plug the pigmented cells indeed form the halves of a neural plate, each accompanied by its neural fold and a tiny notochord (which Roux considered a half-notochord on the

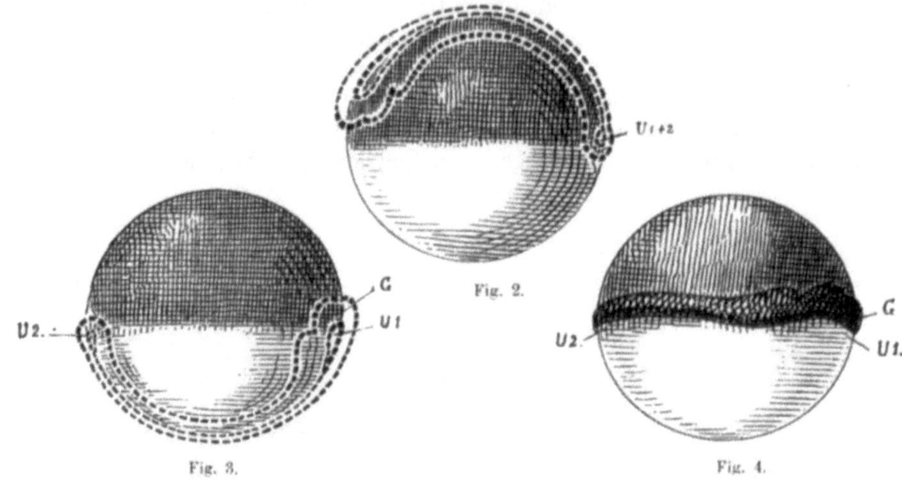

basis of cell counts). The bottom section is from a similar case at a more advanced stage of development, demonstrating that each half-plate develops into a tube and is flanked on its animal side by a single row of somites.

Seeing that the lateral halves of the blocked blastoporal rim develop into the respective halves of the neural plate, Roux concluded (and for the remainder of his life believed) that in normal development, too, the neural plate derived from two separate rudiments localized at opposite sides of the blastoporal lip, joined above the incipient sickle-shaped blastopore by the transverse "brainfold". The lateral rudiments, he thought, grew towards the vegetal midline, where they would meet and fuse like a zipper, beginning in the head region and ending at the tail bud. This, then, is the "bilateral epiboly and concrescence" (or growing together) mentioned in the first quotation. Let it be noted that Oskar Hertwig grudgingly concurred with Roux's interpretation. Indeed, he published diagrams (Hertwig 1892, see next page) showing nicely how the uniform neural plate (*mp*) cut transversely in Fig. 20 is formed by fusion of the two half-plates shown in Fig. 19 – as suggested by the split axis condition that follows from blocked gastrulation (Fig. 17).

The consequences of this view for the inferred orientation of the future embryo in the blastula, and thus for the fate map, were literally revolutionary: the pigmented hemisphere, thought since the days of Karl-Ernst v. Baer to be dorsal, was turned into the prospective ventral side, and the meridian where the blastopore starts forming (U1) changed its prospective value from caudal to cranial, as symbolized in Figs. 2 and 3 on the front page. A glance at the latter figure will reveal that the region antipodal to the incipient blastopore is now indeed ventro-caudal, as claimed in Roux's "Thesis" of 1887 quoted in the previous essay (Sander 1991). As proof of the localization of the brain anlage just above the incipient blastopore (G in Fig. 3), Roux (1888 a)

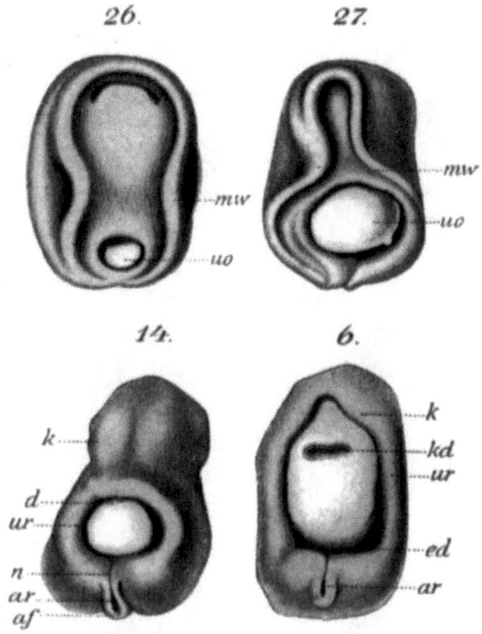

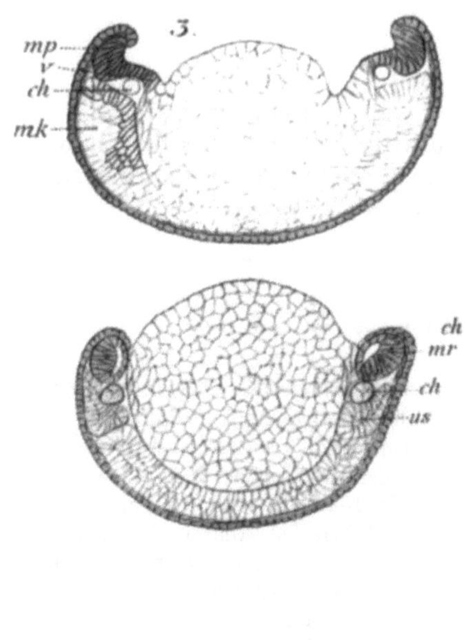

mustered an intriguing result: he had obtained a "formation defect" in the transverse brain fold after destroying the dorsal lip – had he unwittingly destroyed the head-inducing region of the grey crescent by this operation?

Roux's erroneous fate map and mistaken mode of gastrulation embody in their origins an element of irony, for throughout his life (and in opposition to Driesch) Roux proclaimed that the course of development after experimental interference did *not* reveal the mechanisms of normal development.

On the basis of his mistaken views, Wilhelm Roux named the malformation *Asyntaxia medullaris,* that is "failure to grow together in the neural region". Yet he was right in distinguishing this malformation from the *Spina bifida* condition arising by failure of the (uniform) neural plate to transform into the neural tube, a distinction that Hertwig failed to make. By intuition (or just because of his love of terminology?) he suggested a purely descriptive name in addition, *Diastasis medullaris,* which had the advantage of not implying a (wrong) hypothesis (Roux 1888 b).

In closing this essay we shall consider the antecedents and some consequences of the concepts here described. Roux was the first author to capitalize on the blocked gastrula/split axis syndrome in amphibians. However, median clefts in the trunk region had been observed much earlier in teleosts after blocked epiboly, and – granted the homology of blastodisc rim and blastoporal lip – were explained along similar lines (Lereboullet 1863, reviewed in Baumann and Sander 1984). Indeed, what Hertwig called the "Theory of Concrescence" had been proposed by Wilhelm His (1874) on the basis of observations in teleosts. For this reason, the best illustration of the theory (hailing again from Oskar Hertwig, 1892), makes use of fish rather than amphibian development. Hertwig's diagrams 1–4 (on the next page) show nicely how during epiboly the lateral portions (marked ur^2) of the blastodisc rim fuse like a zipper along the future midline (ur^1), thus elongating the axial organs beginning in the head region. If this fusion is blocked, a split axis will result (diagrams 5 and 6). Hertwig, by the way, warmly welcomed Charles Sedgwick Minot (1889), the American anatomist, as yet another (and influential) supporter of the concrescence theory.

As for consequences, it is noteworthy that the replacement of the concrescence theory and its pertinent fate map by differing concepts apparently also sent to oblivion the split axis syndrome that had trig-

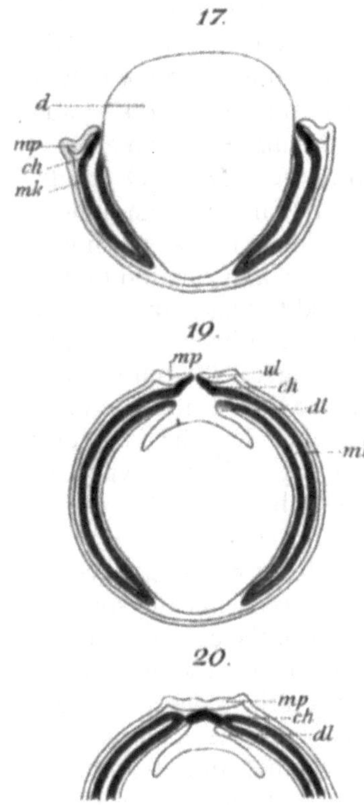

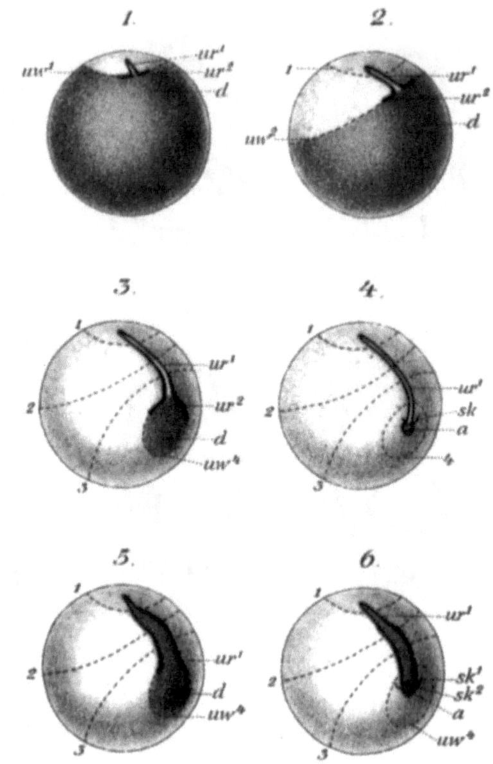

gered the theory. How else could one understand that around 1970 two papers (discussed in Baumann and Sander 1984) independently described the split axis as quite a novel malformation – more than a century after its discovery and essentially correct explanation by Lereboullet? Before the impact of Walter Vogt's fate mapping work (1929), matters were different. Whoever cared was then convinced that in the amphibian fate map the neural anlage abuts the blastoporal rim. Striking proof of this belief is provided by the fact that neither Warren Hammon Lewis (1907) nor Hans Spemann (a decade later, see Sander 1990) was astonished to find neural tissues at sites where they had ectopically implanted pieces of the blastoporal lip – findings that soon afterwards, against the background of the new fate map, were to open the road towards a Nobel prize.

References

Baumann M, Sander K (1984) Bipartite axiation follows incomplete epiboly in zebrafish embryos treated with chemical teratogens. J Exp Zool 203:363–376

Hertwig O (1892) Urmund und Spina bifida. Arch Mikrosk Anat 39:353–502

His W (1874) Unsere Körperform und das physiologische Problem ihrer Entstehung. Vogel, Leipzig

Lereboullet A (1863) Recherches sur les monstruosités du brochet observées dans l'œuf et sur leur mode de production. Ann Sci Natur (Paris) [Sér 4] 20:177–271

Lewis WH (1907) Transplantation of the lips of the blastopore in *Rana palustris*. Am J Anat 7:137–141

Minot CS (1889) The concrescence theory of the vertebrate embryo. Am Nat 24:501–516, 617–629; 702–719

Roux W (1887) Beiträge zur Entwickelungsmechanik des Embryo. 4. Die Richtungsbestimmung der Medianebene des Froschembryo durch die Kopulationsrichtung des Eikernes und des Spermakernes. Arch Mikrosk Anat 29:157–212

Roux W (1888a) Über die Lagerung des Materials des Medullarrohres im gefurchten Froschei. Anat Anz 3:697–705

Roux W (1888b) Beiträge zur Entwickelungsmechanik des Embryo: 5. Über die künstliche Hervorbringung halber Embryonen durch die Zerstörung einer der beiden ersten Furchungszellen, sowie über die Nachentwicklung (Postgeneration) der fehlenden Körperhälfte. Virchows Arch 114:113–153, 246–291

Roux W (1892) Ziele und Wege der Entwickelungsmechanik. Ergeb Anat Entwickelungsgesch 2:415–445

Sander K (1990) From germ plasm theory to synergetic pattern formation: a century of developmental modelling in retrospect (in German). Verh Dtsch Zool Ges 83:133–177

Sander K (1991) Wilhelm Roux on embryonic axes, sperm entry and the grey crescent. Roux's Arch Dev Biol 200:117–119

Vogt W (1929) Gestaltungsanalyse am Amphibienkeim mit örtlicher Vitalfärbung. II. Gastrulation und Mesodermbildung bei Urodelen und Anuren. Roux's Arch Dev Biol 120:385–706

All translations by the present author. The first set of figures is from Roux (1888a); all other figures are from Hertwig (1892). Abbreviations (in literal translation, additions in square brackets): *a* anus; *ar* anal groove; *af* anal folds; *ch* notochord; *d* yolk [mass]; *dl* endodermal lip; *ed* hindgut; *G* brain fold, or rather the site where it will form; *h* adhesion disk; *k* head; *kd* cephalic gut cavity; *mk* mesoderm; *mp* neural plate; *mw* neural folds; *n* fusion line; *sk* tail bud; *U1* site where blastopore starts forming; *U2* site of final blastoporal remnants; *ul* blastoporal lip; *uo* blastopore; *ur* blastoporal rim, ur^1 its parts engaged in fusion, ur^2 the parts situated laterally in the germ disc rim; *us* somite; *uw* edge of epiboly (1–5 at different stages); *v* connecting line [between ectoderm and endoderm]

ESSAY

Landmarks in Developmental Biology (5)

"Mosaic work" *and* "assimilating effects" in embryogenesis: Wilhelm Roux's conclusions after disabling frog blastomeres

Klaus Sander

Wilhelm Roux, after killing individual frog blastomeres, conceived the mosaic theory of embryonic differentiation. At least, that is what the textbooks say. However, this essay will show that Roux claimed self-differentiation (or mosaicism) of blastomeres only for *normal* – in his terminology "typical" – development, and strictly speaking only for its earliest stages, while seeing ample proof of interactive mechanisms in *abnormal* (or "atypical") development after experimental interference. He thus both realized and stated that mechanisms other than mosaic development must exist, and he assumed a role for them in normal development as well; whatever he may have written about mosaic development, he never claimed that this was the only mechanism at work. Indeed, in his first communication on puncturing frog eggs he clearly stated that developmental interaction was required even in self-differentiation: "It should not be necessary to stress again that differentiation, considered as such, is a product of interaction; and that, when distinguishing independent from dependent differentiation, what matters solely is to establish whether the specific cause of a [developmental] change is located within the part undergoing the visible change or outside of it" (Roux 1885, p. 523).

Roux's deletion experiments, according to his accounts (1894, p. 260), were begun in the spawning season of 1882, but it was not until spring 1887 that he punctured single blastomeres of two- or four-cell stage embryos in large numbers. As the spawning season was already drawing to a close, Roux could not avoid a certain incompleteness of the conserved and sectioned material – a shortcoming which gave rise to much doubt and debate later on. He himself (Roux 1888 p. 119) considered his publication to represent no more than a "down-payment" on the problem of self-differentiation; however, further installments by way of experimentation were mainly left to come from other authors, while Roux himself was content to ruminate over theoretical considerations on many subsequent occasions (see e.g. Roux 1895).

Technically, Roux's experiments were fairly simple, as was usual before the advent of the Greenough-type stereomicroscope; the marvellous apparatus developed by Laurent Chabry (1887) for puncturing ascidian eggs under the compound microscope was probably not known to him at the time. So he used a fine dissecting needle to pierce single blastomeres at the two- or four-cell stage, or sometimes two blastomeres at the four-cell stage. He preferred a hot needle fitted with a small brass bead to store heat, and kept it inserted in the blastomere until the surrounding cytoplasm showed signs of coagulation.

Of the many details Roux describes from the development of these embryos, a good part seem haphazard in retrospect, as is typical for pioneering studies, which find it difficult to distinguish the accidental from the essential. As for the latter, the surviving blasto-

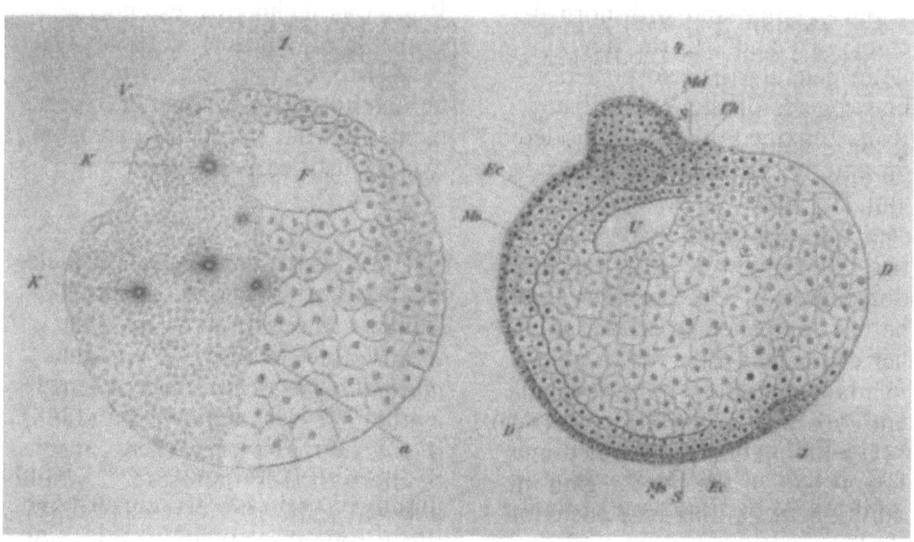

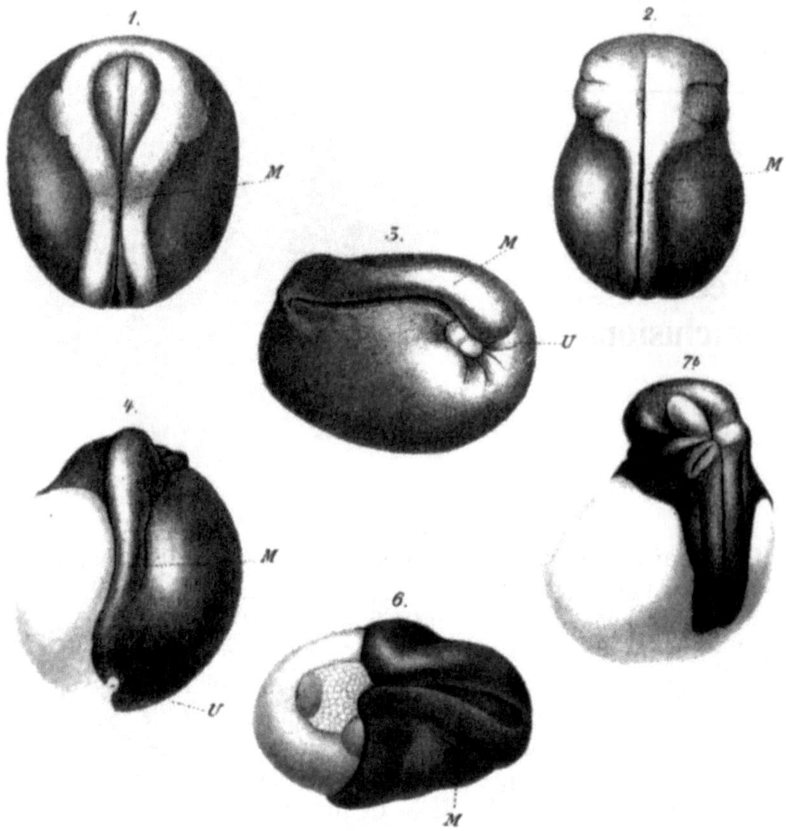

meres from the two-cell embryos continued cleaving, and an asymmetric blastocoele formed (see the accompanying microtome sections). Some time later, the germ layers could be distinguished, which in the most advanced specimens would reveal the anlagen of the notochord, the paraxial mesoderm and half a neural plate with its neural fold (see left hand side of section marked 4). Externally, these embryos could be identified as lateral half-neurulae (embryos marked 3 and 4, compared with normal embryos 1 and 2 in the figure). After deleting the two posterior blastomeres of the four-cell stage, Roux obtained what he considered an anterior half-embryo (embryos 6 and 7b) which in the light of recent findings (see Cooke and Webber 1985) might actually have been dorsoanterior. On the other hand, he felt compelled to retract an earlier claim that after the complementary operation posterior half-embryos would develop – these embryos had actually embodied one lateral half of the head region in addition to the bilateral posterior parts.

Roux summarized his conclusions concisely: "The following principle could therefore be established: the development of the gastrula and of the embryo initially produced from it is, from the four-blastomere stage of the egg on, a mosaic of at least four vertical pieces [the four blastomeres] each developing substantially independently" (Roux 1888 p. 289; translation by H. Laufer, quoted from Willier and Oppenheimer 1964). The words "at least" indicate that Roux was inclined to extend the principle of "mosaic work" (1888, p. 142) to later stages: "I hope to be able to verify by further experiments to what point this extension can be justified."

Turning to the role of the nucleus in mosaic work, Roux assumes – in contrast to his 1883 treatise (see Sander 1991) – that *unequal* segregation starts as early as the *first* cleavage division (1888, p. 138/239). However, there must be more to these mitoses: "I would like here expressly to remark that this [unequal segregation] does not mean a prejudgement as to the distribution of such idioplasm [hereditary material] as functions only in regeneration and postgeneration... A more or less complete idioplasm of this kind is present in every cell, that is to say in each nucleus" (Roux 1888, p. 139, translation by H. Laufer, as above). Roux called this the "reserve idioplasm". He had to introduce it because, much to his amazement, some of the half-embryos (for instance that shown as No. 6) would start to complete themselves by belatedly forming the missing body half, a process he called "postgeneration".

Postgeneration has been regarded with a great deal of scepticism, yet Hans Spemann (1938) for one believed it to be a real occurrence. But even if it were not, the interesting point is the conclusions that Roux drew on its account. He felt that the nuclei in the belatedly forming body half must in many cases derive from the uninjured blastomere – for instance from daughter cells breaking open (*a* in section 1) and despatching their nuclei (*K*) to multiply in the punctured blastomere, which thereafter would be split up by "cellulation" into as many cells. Roux claims that this abnormally generated cell population would segregate into germ layers only in contact with, and starting from, the already well-defined germ layers of the unaffected body half (shown, not very convincingly, in the section marked 4). 'We have to conclude that the causes for this typical continuation of the germ layers of the developed half within the undeveloped half depend on forces which emanate from layers of the developed half... I assume that this differentiation spreads in space by direct assimilatory action of differentiated cells on other, less differentiated cells in their immediate vicinity... This essentially "assimilating" effect of the differentiated cells... must of course be followed by further differentiating effects during the formation of individual organs from these [postgenerated] layers' (Roux 1888, p. 280/281).

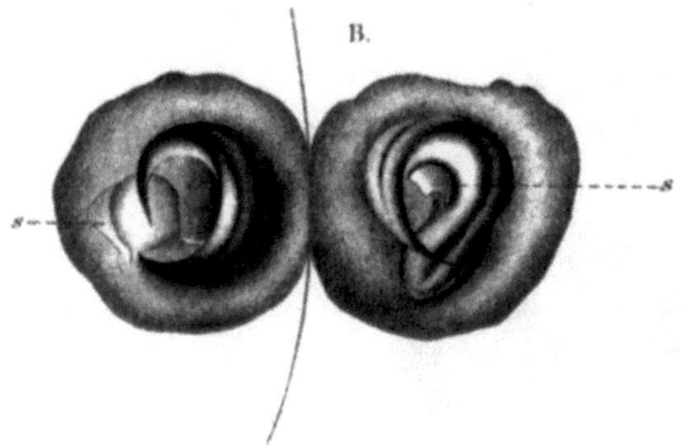

Thus, "the missing half is being postgenerated by dependent differentiaton ... from the primary half that has formed by self-differentiation" (1888, p. 287). In the face of these obviously differing ways to the same end, Roux somewhat surprisingly closes with a credo in the uniformity of causes – which implies dependent differentiation during normal development: "I am convinced that the differences just mentioned, between postgeneration and regeneration, and between both these and normal development, should not be taken to mean that during post- and regeneration essential new formative processes take place which do not occur in normal development. I consider it quite improbable that in the same individual two or three ways of producing the same body parts exist that differ in the essence of their mechanisms; rather one is inclined to presume that postgeneration and regeneration occur under minimal ... deviations from normal development, whereas apart from these deviations the basic events are the same" (Roux 1888, p. 284).

Roux's treatise on half-embryos and postgeneration quickly drew fire from several sides – an early proof for its impact (for references see Spemann 1938). The first volley came (of course) from Oskar Hertwig, who in 1893 presented rather complex evidence showing that single blastomeres from the frog's two-cell stage can directly form complete dwarf embryos. Roux was quick to state that he, too, had seen such embryos, but formed by postgeneration. The shortest, but very enlightening, comments came in 1895 from Thomas Hunt Morgan, who suggested (and to some degree proved) that complete embryos develop when the surviving blastomere is turned upside down, while hemiembryos will result in the absence of turning. His interpretation was sparked by the discovery of Oscar Schultze – published in the first volume of this journal – that turning the normal two-cell stage upside down will frequently lead to partial twinning. That *both* blastomeres of the two-cell stage can produce complete embryos, at least in the newt, was demonstrated in 1895 by H. Endres and soon afterwards by Amedeo Herlitzka, from whose paper the figure on this page is taken. All this, and the subsequent demonstration of "regulative" development by Hans Spemann and his school, laid to rest the idea that early amphibian embryogenesis is mosaic work – except in the mind of Wilhelm Roux, and until experiments in *Xenopus* (Cooke and Webber 1985) showed that isolated frog blastomeres may indeed differ in their developmental fates in a way reminiscent of mosaic development.

References

Chabry L (1887) L'embryologie normale et teratologique des Ascidies simples. Germer Baillière, Paris

Cooke J, Webber JA (1985) Dynamics of the control of body pattern in the development of *Xenopus laevis*: I. Timing and pattern in the development of dorsoanterior and posterior blastomere pairs, isolated at the 4-cell stage. J Embryol Exp Morphol 88:85–112

Herlitzka A (1897) Sullo sviluppo di embrioni completi da blastomeri isolati di uova di tritone (Molge cristata). Arch Entwicklungsmech Org [Roux'Arch] 4:624-658

Roux W (1885) Beiträge zur Entwickelungsmechanik des Embryo: 1. Zur Orientierung über einige Probleme der embryonalen Entwicklung. Z Biol 21: 411–524

Roux W (1888) Beiträge zur Entwickelungsmechanik des Embryo: 5. Über die künstliche Hervorbringung halber Embryonen durch die Zerstörung einer der beiden ersten Furchungszellen, sowie über die Nachentwicklung (Postgeneration) der fehlenden Körperhälfte. Virchows Arch 114:113–153, 246–291

Roux W (1894) Die Methoden zur Hervorbringung halber Froschembryonen und zum Nachweis der Beziehung der ersten Furchungsebene des Froscheies zur Medianebene des Embryo. Anat Anz 9:248–262, 265–283

Roux W (1895) Gesammelte Abhandlungen über Entwickelungsmechanik der Organismen (2 vols, 816 and 1075 pp). Wilhelm Engelmann, Leipzig

Sander K (1991) Wilhelm Roux's treatise on "qualitative" mitoses – a "classic" by either definition. Roux's Arch Dev Biol 200:61–63

Spemann H (1938) Embryonic development and induction. Yale University Press, New Haven

Willier BJ, Oppenheimer JM (eds) (1964) Foundations of experimental embryology. Prentice-Hall, Englewood Cliffs, NJ

Translations not marked otherwise are by the present author, with added explanations set in square brackets. The figures showing frog embryos are from Roux 1888 (the external views are differently labelled reprints of 1895), and the newt twins were published by Herlitzka (1897).

Abbreviations: a, open cell; *Ch,* notochord; *D,* yolk cells; *Ec,* ectoblast; *F,* blastocoel; *J,* yolk cells that remained juvenile; *K,* nucleus; *M, Md,* neural fold; *Ms,* mesoblast; *s,* extraovate [result of intentional severing of the blastomeres]; *S-S,* median plane; *U,* archenteron [in sections], blastopore [in embryos]; *V,* vacuoles.

ESSAY

Landmarks
in Developmental Biology (6)

Wilhelm Roux and the rest: Developmental theories 1885–1895

Klaus Sander

This last essay in the series commemorating Wilhelm Roux (1850–1924) is devoted to speculative developmental theories (or models as they would be called today) proposed in the decade following Roux's manifesto on *Entwickelungsmechanik* (see Sander 1991a). That decade, 1885–1895, not only saw Roux's pioneering experimental papers discussed in previous essays and the influential books of August Weismann (1834–1914) on his *Keimplasmatheorie*, but also the competing theories on animal development proposed by Hans Driesch (1867–1941) and Oscar Hertwig (1849–1922). Towards the end of this period, Roux founded his *Archiv für Entwickelungsmechanik,* demonstrating the new discipline's consolidation.

That decade was singular in the wealth of speculative developmental theories it spawned. They were evidently influenced by the new understanding of mitosis and the role of the nucleus (see Sander 1991b). After the decade, further far-flung speculation may have been discouraged by the complex results of experimental research stimulated by those theories. In this context it seems only natural that the same decade embraced the gestation period of Edmund B. Wilson's magnificent "Cell" (1896), the book which more than any other heralded modern developmental biology; Wilson was indeed the first author to show (in 1893) how the conflicting theories of the decade could be integrated in a single concept. He dedicated his book to Theodor Boveri (1863–1915), who by combining karyological and experimental techniques had introduced promising alternatives to mere speculation.

August Weismann shared Roux's mechanistic and reductionist attitude, which is documented both in the previous essays and by Roux's writings before and after the decade under review (see Oppenheimer 1965; a list of Roux's publications is found in *Die Naturwissenschaften* 8:457–459, 1920). But whereas for Roux ontogenesis was the central concern, Weismann's speculations on ontogenetic mechanisms were part of a grandiose attempt to understand Darwinian evolution on the basis of the karyological knowledge that had accumulated since the "Origin of Species" (see e.g. Mayr 1982). Weismann envisioned the progressive splitting-up of the hereditary substance by unequal mitoses during development. He illustrated this concept (Weismann 1892) using an "infinitely condensed" fictive cell lineage of the forelimb (see figure) whose "real connection in any case must be quite different" (but re-

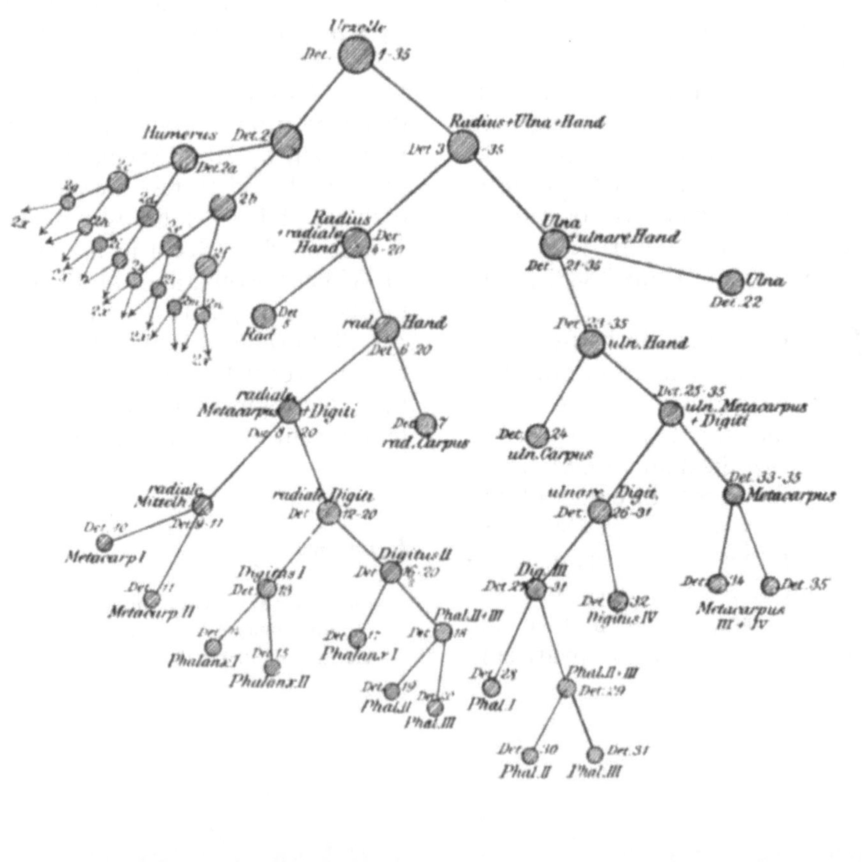

mained unresolved). The founder cell contains in its nucleus solely those determinants (numbered 1–35) that specify the different skeletal elements, and these determinants are then parcelled out during successive mitoses until each cell is left with only a single determinant that directs the development of its progeny.

Evidently this scheme involves unequal mitoses, as proposed for early cleavage by Roux, but there the similarities end. For Weismann implies that all these unequal divisions and also the spatial relations between the respective daughter cells are encoded solely in the molecular architecture of the nuclear germ plasm itself, whereas Roux entrusts the anisotropic cytoplasm (or "yolk") of the dividing cell with the task of properly orienting the spindle and thus locating the unequal daughter nuclei: "So, if mitosis is capable of segregating the [nuclear] qualities properly, ... there is the additional requirement that the products of segregation are brought into the appropriate [spatial] relations to the different parts of the [dividing] cell's body and to neighbouring cells already present" (Roux 1885b, p. 33). In explaining how this requirement could be met, Roux refers to a spindle aligned (as in second cleavage) parallel to the egg cell's median plane and equator: "In this case one end of the spindle will point to that side of the egg where the white cap reaches higher up [forming the grey crescent], the other end towards the darker side of the cell's body; and the respective differences in the substances of the cell's body cause that the nucleus correspondingly divides in qualitatively unequal fashion ..." (Roux 1893, p. 148). Moreover, Roux was convinced that a vast amount of cellular interaction would be required for later stages of development (such as depicted in Weismann's sketch): "The multitude of qualitative and formally differentiating as well as ordering ... interactions of cells ... we cannot as yet estimate; on these [however] depends predominantly the shaping of the multicellular organism and especially its integrity" (Roux 1897, p. 306). The points that Roux made against Weismann's autonomous segregation of determinants are well taken: "Such a rigid order would exclude any ‹self-regulation›; and by a single error the entire series of subsequent divisions would be directed onto a wrong track" (Roux 1885b, p. 34).

The concepts of Weismann and Roux of course were both fraught with the problems posed by regeneration and by postgeneration (or whatever process produced whole embryos from isolated blastomeres). They forced Roux (and similarly Weismann) to assume that many cells receive a complete "reserve idioplasm" during mitosis, in addition to their share of specific determinants. The reserve idioplasm remains dormant until it is "activated" by situations that require "atypical development" for re- or postgeneration, whereas the germ cell idioplasm that splits up during "typical development" is activated, in Roux's scheme, by fertilization.

This cumbersome duplicity was eliminated by the other pair of theories we shall now discuss. These – equally speculative – theories were put forward by Driesch and Oscar Hertwig (Oscar as distinct from his brother Richard, another eminent zoologist). They maintained that every somatic cell contains the complete idioplasm of the germ cell, and this not just for emergency repairs but primarily for normal development. Their claim was supported by the novel finding (Driesch 1892a, actually issued 10.11.1891) that separation of sea urchin blastomeres can lead to complete dwarf larvae (a_3 and b_3 in the figure) rather than to hemiembryos (a_2 and b_2, as opposed to the normal embryos a_1 and b_1). Hertwig argued (correctly) that Roux's half-embryos might owe their existence to effects connected with the adjacent disabled blastomere. Both authors maintained that pattern formation occurs epigenetically by complex interactions between the many cells generated by cleavage. "The parts of the organism develop in relation to each other, that is, the development of a part is dependent on the development of the whole" (Hertwig 1892, p. 480). Hence a cell's developmental fate is "a function of its position" (Driesch 1892b, p. 39) rather than of its descent in the lineage.

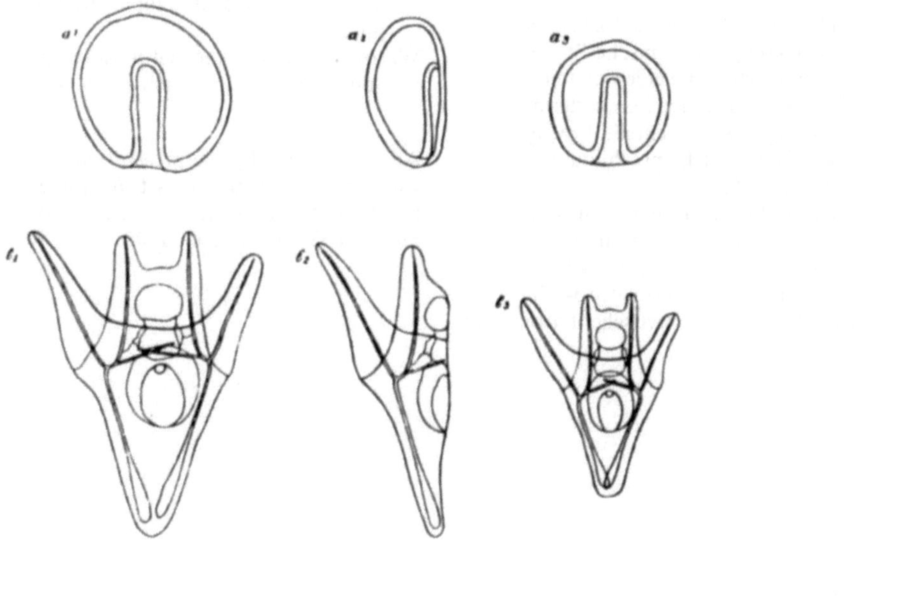

Driesch subsumed both Hertwig's and his own theories under the name "Theory of Epigenetic Evolution" – evolution being, in the usage of the period, the generation of organismic shape from a precursor structure preformed invisibly in the egg (for our authors, the idioplasm in the nucleus). But this name would have applied equally to the sum of Roux's ideas for, as we have seen, he also envisioned the body arising by epigenetic interactions directing the expression of genetically preformed information (which, however, he expected to split up during the process).

The heuristic value of these early theories is evident from the course that developmental biology took thereafter. In their factual content, the ideas of Driesch and Hertwig seem closer to what we know or believe today, but Roux's speculations were the most influential, having triggered those of Driesch and Hertwig to begin with. Interestingly, Weismann had also (and earlier) considered the possibility that all cells contain the complete germ plasm, only to reject it with a very revealing argument: "Why should Nature, who lets parsimony act everywhere, afford the luxury to endow all cells with all the determinants in the germ plasm if a single kind suffices?" (Weismann 1892, p. 85). Behind this argument stands of course the same staunch belief in Darwinian selection that pervades the writings of Roux. Driesch and Hertwig, on the other hand, both rejected Darwin's principle of natural selection: Driesch, in 1896, called this principle "the curiosity of our century", and during the First World War Hertwig published a manifesto on "*Warding off Biological, Social and Political Darwinism*".

With these ideological differences and the fact that speculative theories invite heated arguments, it is no wonder that their protagonists – except for the ever gentle Weismann – tended to argue *ad personam*. Roux, himself not innocent in this respect, felt the urge to admonish prospective contributors to his newly founded *Archiv*: "But one limitation is desirable in the forthcoming struggle of opinions ...: observing a respectful tone towards the exponents of opinions deviating from one's own. The search for truth, which is our common aim, will not be furthered but rather impeded by the expression of personal feelings" (Roux 1895, p. 37).

These admonitions may not always have been heeded, but in later years even Roux's most ardent and eloquent ideological opponents abided by them. Thus, Hans Driesch wrote in the historical introduction to his vitalistic magnum opus (1909, Vol. I, p. 57): "I cannot let pass this occasion without expressing in the strongest ways how highly must be valued, in my opinion, the importance of Roux for the systematic exploration of the generation of [biological] form. I feel all the more obliged to do so because I shall later on be compelled to contradict not only many of his results but also most of his theoretical considerations ... But Roux's place of honour in the history of biology is certain, may this science take whatever course it will". Which has proven true.

Acknowledgements. The author wishes to express his thanks to Prof. J.H. Scharf (Halle/Saale) for providing the photograph of the medal commemorating Roux, to Gudrun Mahlke for photographic reproductions, to Margrit Scherer for untiring editorial assistance, and last but not least to the staff of Springer-Verlag for generous support.

References

Driesch H (1892a; issued 10.11.1891) Entwicklungsmechanische Studien: I. Der Werth der beiden ersten Furchungszellen in der Echinodermenentwicklung. Experimentelle Erzeugung von Theil- und Doppelbildungen. Z Wiss Zool 53:160–178

Driesch H (1892b) Entwicklungsmechanische Studien VI: Über einige allgemeine Fragen der theoretischen Morphologie. Z Wiss Zool 55:34–58

Driesch H (1894) Analytische Theorie der organischen Entwicklung, Engelmann, Leipzig

Driesch H (1909) Philosophie des Organischen (Gifford Lectures). Engelmann, Leipzig (English edition 1908)

Hertwig O (1892) Urmund und Spina bifida. Arch Mikrosk Anat 39:353–503

Mayr E (1982) The growth of biological thought. Belknap/Harvard University Press, Cambridge / Mass

Oppenheimer JM (1965) Questions posed by classical descriptive and experimental embryology. In: Moore JA (ed) Ideas in modern biology. Natural History Press, Garden City / NY, pp 205–227

Roux W (1885a) Beiträge zur Entwickelungsmechanik des Embryo: I. Einleitung. Z Biol 21:411–524 (reprinted in Gesammelte Abhandlungen vol 2)

Roux W (1885b) Beiträge zur Entwickelungsmechanik des Embryo: III. Ueber die Bestimmung der Hauptrichtungen des Froschembryo im Ei und über die erste Theilung des Froscheies. Breslauer Ärztl Z 1885 Nr. 6ff:1–55 (reprinted as above)

Roux W (1893) (Discussion on the talks of O. Schultze and H.E. Ziegler). Verh Anat Ges 1894:148 (reprinted as above)

Roux W (1895) Einleitung. Arch Entwikkelungsmech Org 1:1–42

Roux W (1897) Für unser Programm und seine Verwirklichung. Arch Entwickelungsmech Org 5:1–80, 219–342

Sander K (1991a) Wilhelm Roux and his programme for developmental biology. Roux's Arch Dev Biol 200:1–3

Sander K (1991b) Wilhelm Roux's treatise on "qualitative" mitoses – a classic by either definition. Roux's Arch Dev Biol 200:61–63

Weismann A (1885) Die Continuität des Keimplasmas als Grundlage einer Theorie der Vererbung. Fischer, Jena

Weismann A (1892) Das Keimplasma: Eine Theorie der Vererbung. Fischer, Jena (English edition 1893)

Wilson EB (1983) Amphioxus, and the mosaic theory of development. J Morphol 8:579–617

Wilson EB (1896) The cell in development and inheritance. Macmillan, New York

All translations by the present author, with added explanations set in square brackets. The illustrations are from Weismann 1892 and Driesch 1909.

ESSAY

Landmarks in Developmental Biology (7)

Laurent Chabry (1855–1893) and the "Age d'or" of biology

Jean-Louis Fischer

Editor's note: The series of historical essays initiated in Vol. 200 of Roux's Archives will be continued. While the general aim is to discuss classical data or methods in their historical context, the present essay is mainly biographical, because the author has recently discovered previously unknown documents, among them the first known portraits of Chabry.

When, towards the end of his life, Eugène Bataillon (1864–1953) remembered the period around the turn of century, which was so rich in milestones of theoretical biology, he considered it the "âge d'or de la biologie" (Bataillon 1953, p. 384). He himself, the discoverer of traumatic parthenogenesis in the frog (1910) and an expert in the theory of fertilization, had made a substantial contribution to the image of biology in those years, owing to his elaboration of extremely refined methods of experimenting on the egg. The reasons for the embryologist's invention of these experimental methods and the theoretical conclusions resulting from their application both originated in the 1880s. During those years comparative embryology, aiming at the genealogy of species, gave way to experimental embryology that disclosed cellular genealogies – the embryologists started moving from evolutionary to developmental problems. Laurent Chabry was among the scientists who forged that new biology, the "*science nouvelle*" that originated from experimental embryology, named *biomécanique* by Yves Delage (1854–1920) in 1895 and previously conceived by Wilhelm Roux (1850–1924) as *Entwickelungsmechanik* in 1884 (Sander 1991).

Few facts are known about the life of Laurent Chabry, and some of them (and the portraits printed here for the first time) came to light only a few months ago. Chabry was born on 19 February 1855 at Roanne in the département de la Loire. Young Laurent is said to have possessed *aptitudes remarquables*. In 1876 he spent several months in St. Petersburg in Russia to perfect his knowledge of Russian, which enabled him to translate scientific papers (e.g. those published by A. Kowalewsky, see Figure on the next page) from that language. After his return to France he became active in politics, supporting the *collectivisme révolutionnaire* called into life by Jules Guesde (1845–1922). He took part in organizing the 1878 International Workers' Congress. This congress, planned to take place during the World Exhibition at Paris, was forbidden by the authorities. As the organizers went on with it nonetheless, Chabry, together with the other members of the organizing committee, was taken into custody by the police and committed to prison (Haupt and Verdès 1965). In this connection Maurice Caullery (1868–1958) wrote: "This circumstance to me seems worth remembering, in order to characterize his passionate personality and also a certain restlessness of spirit which made him depart from the common and easy ways" (Caullery 1940, p. 230). In 1877, Guesde had founded a weekly journal of international ambitions, *l'Égalité*, to publicize his collectivist ideas. Chabry became a member of the administration committee, and we find his signature at the end of a manifesto that the *Socialistes Revolutionnaires Français* addressed to those attending the 8th Congress of the German Social Democrats: "Brothers in Germany, the revolutionary socialists of France do not wish to let your eighth congress open without assuring you of their sympathies and their fraternal support" (*l'Egalité*, June 16th, 1878).

Besides his political activities, while in Paris Chabry embarked on the study of medicine and in 1881 submitted his doctoral thesis, entitled *Contribution à l'étude du mouvement des côtes et du sternum*. This topic may seem far from his subsequent preoccupation with embryology, but there is a certain intellectual logic in the shift from the mechanics of organs to the mechanics of embryogenesis.

Chabry entered the life sciences with a particularly favourable background, namely his good knowledge of European languages –

in the early 1880s German became the scientific language that was indispensable for those planning to work in physiology or embryology – and also his mathematical training and his knowledge in chemistry and physics. His first inclination was to be a physiologist. Etienne Jules Marey (1853–1904), whose lectures he attended at the *Collège de France*, exerted a great influence on him, as is revealed by the subject of his medical thesis and by his investigations on the mechanics of jumping and flying (Fischer 1990, 1991).

To begin with, Chabry worked for some time at the *Sorbonne* in the zoological laboratory of Henri Lacaze-Duthiers (1821–1901), and it was the contacts with this great zoologist that made Chabry turn to marine biology. We owe to Lacaze-Duthiers the installation of the marine stations of Roscoff (1872) and Banyuls (1881). Two other scientists were to complete the biological education of Chabry: Charles Robin (1821–1885) and Georges Pouchet (1833–1894). Both were convinced positivists. Robin, the founder of the *Societé de Biologie* (1849), a specialist in animal histology and renowned encyclopaedist, was the teacher of Pouchet. The latter, after starting in the field of anthropology, had turned to histology and developed a passion for marine biology. He enjoyed worldwide recognition as a specialist on whales (Pettit 1902). He held the chair of comparative anatomy in the *Muséum d'histoire naturelle de Paris* from 1879 to 1894. After the 1870 war, Pouchet founded a histological laboratory in Paris at his own expense, which in 1872 was merged with the Laboratory of Histological Zoology of the *Ecole des Hautes Etudes* headed by Robin. Several years later, Chabry was to spend some time as an "assistant" in this very laboratory.

All three scientific personalities who, from the 1880s onward, were to guide Chabry's first steps in the life sciences – Marey, Robin and Pouchet – were medical men like Chabry himself. They all had a mechanical concept of the phenomena of life: movement is responsible for form, the milieu exerts preponderant influences on movement and consequently on form, the function creates the organ and adaptation is conceived as a mechanism of the evolution of form. All told, these are the grand ideas of Neo-Lamarckism, which they expected to be confirmed by experimentation and which were so different from the néo-darwinism of the *fin du siècle*.

But beyond these tenets of a scientific ideology that was important for Chabry's choice of theories, there were people who were specifically trying to understand the developmental mechanics rising to prominence in the 1880s. The origins of this new embryological discipline lay with several biologists, one of whom – although not always remembered – is Marey, as related by Alfred Giard (1846–1908): "Marey must be considered the precursor of the modern school of morphodynamic embryogenesis that flowers so much in Germany" (Giard 1905, p. 5; for a bibliographic study of Marey see François-Franck 1905; Pouchet 1892). Indeed, in a synthesis of his ideas, which were close to the organicism of W. Roux but had already been proposed in *La machine animale* (1872), Marey wrote: "The field of these experiments is unlimited, so to speak, and therefore I recommend this subject to

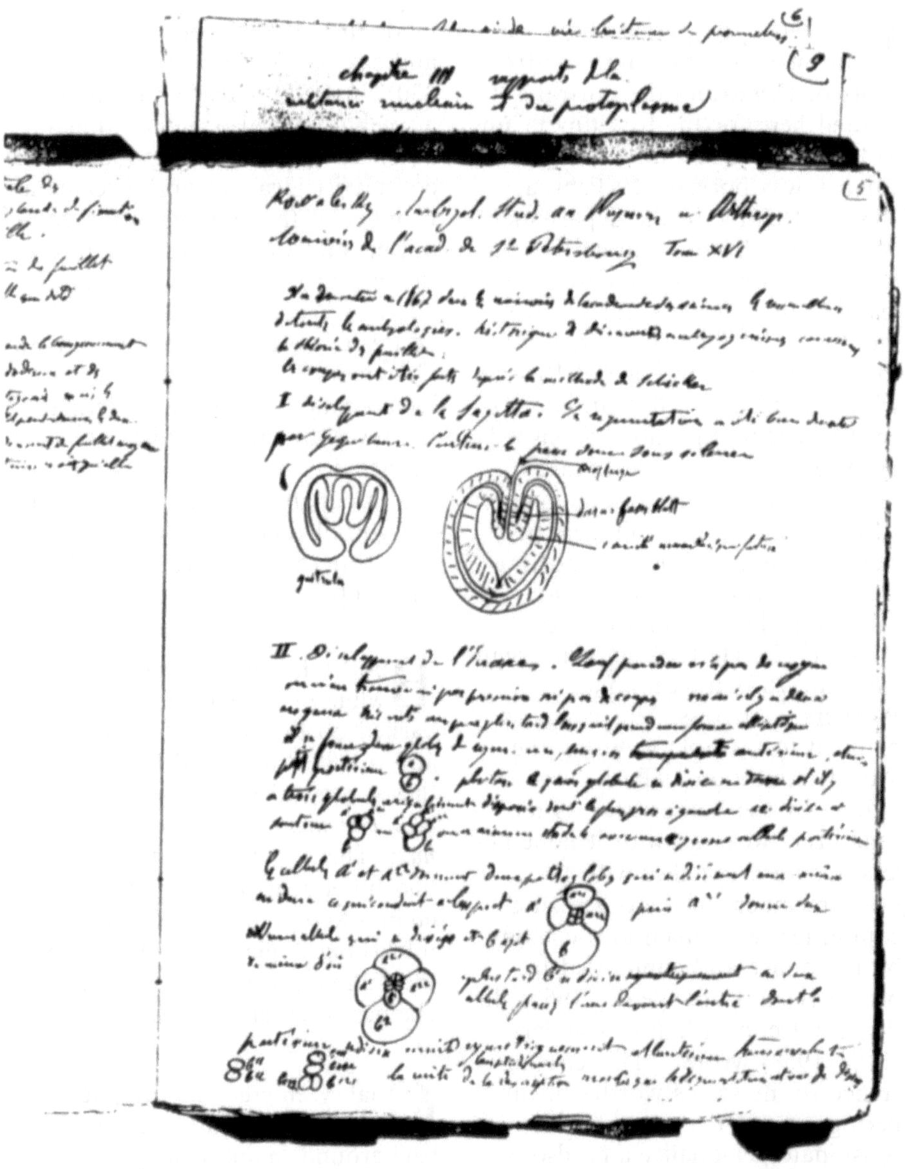

the attention of the researchers and it will be the task of the young generation to develop this new branch of science, the *morphogénie expérimentale*" (Marey 1889, p. 100). Marey wrote these lines after Chabry had defended his biological Ph.D. thesis, but they reveal a scientific fraternity sharing ideas, and a common guideline accepted in biological research at that time.

After his term at the *Ecole des hautes études*, Chabry took the position of assistant director in the marine biological laboratory at Concarneau, the first marine station, founded in 1859 by Victor Coste (1807–1873). Curiously, according to Pouchet, Chabry refused an offer from Marey to work in his new laboratory, which the *Collège de France* was to establish in 1880: la *Station physiologique du Parc des Princes*. His choice to move to the Concarneau laboratory might be explained by the fact that Robin was the director there, and Pouchet the co-director. Thus, the integrity of the trio of scientists from the *Laboratoire d'histologie* was maintained. The choice may also have been influenced by Chabry's interest in marine biology, kindled by Pouchet.

From 1880 to 1884 Chabry published several scientific notes in collaboration with Pouchet or Robin. He studied the origins of enamel and the odontology of mammalians as well as the fetal stages of the dolphin, and he continued his studies in animal mechanics (swimming mechanisms in fish, jumping mechanisms, etc.). In 1884 he started to work in the field of "mécanique cellulaire", which Pouchet had defined as follows: "We consider *mécanique moléculaire* very much at the core of the living substance, in its relation to the morphology of living beings" (Pouchet 1893, p. 736). Chabry's material of study was the ascidian larva and this research enabled him to submit his thesis in the natural sciences in 1887; this was entitled *Embryologie normale et tératologique des Ascidies* and is recognized as one of the pioneering studies in causal embryology.

Accepted as a member by the *Société de Biologie* in April 1887, Chabry hoped that after defending his thesis he would obtain a chair of zoology in the faculty of sciences at Lille. Meanwhile, he took up the post of *maître de conférences* in the faculty of sciences at Lyon, in the division of zoology directed by Professor Sicard. It was there that he met Bataillon, who was *préparateur* in zoology. Chabry exerted a profound influence on Bataillon by his technical talent, his ability and his *savoir faire*. When Bataillon discovered traumatic parthenogenesis in 1910, he remembered that it was Chabry who had introduced him to the technique of producing microneedles suitable for precision operations on the egg (Bataillon 1955; Rostand 1959; Fischer and Smith 1984). Little satisfied with his situation in Lyons, and somewhat disappointed because he had not obtained anything better than this post, in 1890, after two years of service, Chabry submitted his resignation to the faculty of sciences in order to devote himself to dentistry and facial surgery. He settled in Paris and set up in practice there.

Thus, Chabry's time in developmental biology came to an end. Pouchet thought that the educational authorities had made a mistake by not keeping this exceptionally valuable researcher in service. Caullery, however, did not share Pouchet's judgement, because in his view Chabry himself was responsible for the mistake. He felt that Chabry was lacking in patience, because in 1892 the chair of the department to which he had been attached would of course have been offered to him after falling vacant: "Having proven himself an eminent biologist and seeing his value acknowledged, he – who was so engaged for the idea of social justice should have noticed the advantages that science has to offer. It was obvious that the chair of zoology at Lyon, which could not be given to him immediately, would be vacant within a short time and, in fact, it became so two years later, in 1892" (Caullery 1940, p. 232). In any case it is true that Chabry was the

strong character characterized by Rostand: "*Original, capricieux, indépendant jusqu'à en être rebelle...*" (Rostand 1959, p. 4). The portrait by an unknown artist shown here seems to reflect all these traits.

But Chabry could not long do without the atmosphere of the research laboratory, being in the words of Elie Metchnikoff (1875–1916) – "*pardessus tout l'amour désintéressé de la science*" (Metchnikoff, address delivered at Chabry's funeral). Therefore, he accepted the position offered him by Metchnikoff to do research in bacteriology at the *Institut Pasteur*.

There he undertook work which, according to the testimony of Metchnikoff, was about to lead to promising results: "His independent character thrived at that liberal institution, which gave full freedom to his investigative spirit. There he valiantly took on a project likely to occupy him for some years. In this he poured that ingenuity and that experimental ability which were his specific qualities, and as he showed a tenacity that never tired it is certain that his bacteriological investigations would not have been unworthy of those that had marked his entrance into biology" (Metchnikoff 1893).

While handling the *bacilles de Koch* during his experiments,

Chabry was infected by the microbe. A journalist writing in *Le Mémorial de la Loire et de la Haute-Loire* (26 and 28 November, 1893) quoted Chabry as saying: "I have been infected with the bacillus and shall die. This is not astonishing; after my innumerable manipulations I permanently had tuberculous matter at the tips of my fingers, and you know my habit of pulling my mustache. I thus have swallowed the germs of disease." Being a physician turned "pasteurian" and an old pupil of the renowned hygienist Marey – who in his report on the cholera epidemy of 1884 had concluded that "it is the contamination of the water used for public consumption" that is responsible for transmitting this lethal disease – it seems nearly inconceivable that Chabry could have committed such a blunder. Yet he did. Chabry was himself responsible for his infection by the Koch bacillus, which led to the renal tuberculosis that proved fatal to him (for symptoms, see Pouchet 1893, p. 738). In the words of the *Journal de Roanne*, this death (on 30 November 1892) made him "*un martyr de la science.*"

After becoming ill he had to leave the Institut Pasteur, hardly 18 months after joining it. He went to stay with his brother at Riorges in search of recovery. He continued reflecting on the mechanics of animals and pondered the notion of a flying machine based on the propeller. But his powers left him and he died on 23 November 1893, aged 38. In his will he left 5000 francs to the laboratory at Concarneau and 5000 francs to the *Institut Pasteur*; this was all he had owned.

The journals of the Roanne region reported that Chabry had converted to the Christian faith. The possibility that this "wild socialist" might have reneged on the ideas he had previously defended did not fail to provoke an allusive criticism from Pouchet. "From the first day of his disease, or rather from his retreat from Paris, and even when he did not yet believe that he was seriously affected, the friends of Chabry did not receive any news from him – he anyhow wrote very little – but there was no doubt among them that their old comrade, their ancient *compagnon d'armes* in the battle of ideas, had not lost, as long as his intelligence was not affected, his scientific creed" (Pouchet 1893, p. 738).

Chabry was sadly missed in the domain of the life sciences. By virtue of his great technical ingenuity and his ease in resolving delicate problems with an important bearing on theory, his embryological work was sufficient to earn him an international reputation, and he is justly recognized as one of the pioneers of experimental embryology.

Acknowledgements. For information and documents concerning Laurent Chabry we wish to thank Mr. Legal, director of the Laboratoire de biologie marine du Collège de France, Miss Rochet-Benétière, Mr. Denys Barau from the *Archives de la Loire* and Mr. Norbert Chabry. The originals of the portraits and the manuscript shown here are in the archives of the Laboratoire de biologie marine du Collège de France at Concarneau. This essay was translated from the French original by K. Sander.

References

Bataillon E (1953) "Hommages à Jean Rostand". In: Instruire sur l'homme. La Diane Française, Nice

Bataillon E (1955) Enquête de trente cinq ans sur la génération. Sedes, Paris

Brachet A (1921) Traité d'embryologie des vertébrés. Masson, Paris

Caullery M (1939) Les progrès récents de l'embryologie expérimentales. Flammarion, Paris

Caullery M (1940) Laurent Chabry (1855–1893). Rev Sci 78:230–232

Delage Y (1895) Une science nouvelle: la biomécanique. Rev Gén Sci Pures Appl 6:441–446

Fischer JL (1990) Experimental embryology in France (1887–1936). Int J Dev Biol 34:11–23

Fischer JL (1991) Laurent Chabry and the beginnings of experimental embryology in France (translated by M. Krausz and Scott F. Gilbert). In: Developmental biology, a comprehensive synthesis, vol 7: Gilbert SF (ed) A conceptual history of modern embryology. Plenum Press, New York, pp 31–41

Fischer JL, Smith J (1984) French embryology and the "mechanics of development" from 1887 to 1910: L. Chabry, Y. Delage and E. Bataillon. Hist Philos Life Sci 6:25–39

François-Franck M (1905) Biographie du professeur E.J. Marey, ancien Président de la Sociéte de Biologie. C R Soc Biol 59:1–22

Giard A (1905) Allocution présidentielle. C R Soc Biol 58:4–7

Haupt G, Verdès J (1965) De la première à la deuxième internationale, les actes des congrès internationaux, 1877–1888: répertoire. Le mouvement social 51:113–125

Marey J (1889) Des lois de la morphogénie chez les animaux. Arch Physiol Norm Pathol 21:88–100

Pettit A (1902) Georges Pouchet (1833–1894) Notice biographique in Recueil des principales oeuvres de Ch. H. Pouchet. Masson, Paris

Pouchet G (1892) La forme et la vie. Revue des deux mondes (15 May)

Pouchet G (1893) Laurent Chabry. J Anat Physiol Homme Anim 29:735–739

Rostand J (1959) Hommage à Eugène Bataillon. Paper presented at the 84th Congress of the Sociétés Savantes de Paris et des départements à Dijon, Science section

Roux W (1884) Beiträge zur Entwickelungsmechanik des Embryo: II. Über die Entwickelung der Froscheier bei Aufhebung der richtenden Wirkung der Schwere. Breslauer Ärztl Z 22:1–16

Sander K (1991) Landmarks in developmental biology. Wilhelm Roux and his programme for developmental biology. Roux's Arch Dev Biol 200:1–6

ESSAY

Landmarks in Developmental Biology (8)

The embryological œuvre of Laurent Chabry

Jean-Louis Fischer

The embryological work of Laurent Chabry (1855–1893) consists of his doctoral thesis, *Embryologie normale et tératologique des Ascidies simples* (1887, see next page), supplemented by some research on the sea urchin egg (1886–1889). At the *laboratoire de zoologie maritime* of Concarneau and in parallel with his work on animal mechanics (see Fischer 1992), in 1884 Chabry started a series of investigations on *mécanique cellulaire.* To allow him to observe the egg of *Ascidia aspersa*, he invented the *capillaire porte-objet*, to which he then added a *perforateur* for his experiments (see next Essay). His technical experience, backed up by his exceptional abilities, permitted him to elucidate developmental facts with the aid of appropriate instrumentation.

In the history of the experimental method in embryology, Chabry was the first to level a direct attack at the individual cell or blastomere under the microscope. In doing this, he removed an epistemological barrier and opened the way to a substantial series of experiments which, by way of their theoretical implications, were to provoke a debate on neo-preformation versus neo-epigenesis that was absolutely fundamental to the history of biology.

When Chabry began his descriptive studies on the origins of blastomeres and on the gastrula and first larval stages of *Ascidia*, he continually encountered batches of eggs with developmental anomalies that led to the formation of monstrous larvae. As this annoying circumstance prevented him from continuing his studies of normal development, he decided to exploit the abnormally cleaving eggs. It was because of this novel and unforeseen turn in his research programme that Chabry came to study *un sujet absolument neuf*. He observed and enumerated seven types of cleavage anomaly: (1) deviation of a cleavage plane, (2) delayed cleavage, (3) cleavage restricted to the nucleus, (4) absence of cleavage, (5) fusion of blastomeres, (6) abnormal movement of blastomeres, and (7) death of one or several blastomeres. These seven processes, which he called *hémitériques* (monstrous), were rarely met with singly; much more frequently several of them coexisted in a single egg.

Here we shall restrict ourselves to the observations that Chabry made with respect to the death of one or several blastomeres. At the two-blastomere stage, the death of one blastomere led to the development of what Chabry considered a left or right half-individual. The death of one or two blastomeres at the four-blastomere stage led to the formation of three quarters or two quarters of an individual. These observations determined his choice of theory for explaining ascidian embryogenesis: it was the theory of preformation. This choice prompted him to devise a plan for experimentation.

By the time of these descriptive investigations (1884), quite a number of biologists had performed embryological experiments to produce monstrosities. The most famous was Camille Dareste (1822–1899; see Fischer 1987, 1991) who, beginning in 1855, had systematically produced monstrous chick embryos (Dareste 1891); he was influenced by Etienne Geoffroy Saint-Hilaire, who in the 1820s had made some timid attempts at experimental teratology. Dareste used the *methode indirecte*: he exposed the entire egg to some teratogenic factor, for instance lowered incubation temperatures for several hours during early development. The aim that guided Dareste in his work was the achievement of some *transformisme expérimental* of breeds or even species, and when he reflected on the embryological aspects this was clearly not in keeping with his prime preoccupation: Dareste was akin, long before their time, to the early twentieth century mutationists who would work in the new scientific context of genetics – he really does not belong among the *nouveaux embryologistes* of the 1880s and 1890s. Auguste Lereboullet (1864) practised experimental teratology in fish eggs for some time, but he did not believe in the potential of experimentation to produce monsters. Next, the Swiss physiologists Stanislas Warynski and Hermann Fol (1884) inaugurated the *méthode directe* in their experimental approach: with a thermocauter they destroyed part of the chick embryo, for instance the head rudiment, at 24 or 48 hours of incubation. Their objective was to generate an anomaly in the embryo in order to compare this with normal development. They obtained, among other anomalies, anencephalies and duplications of the heart. They could reproduce these monstrosities at will, but they too failed to extend their initial work. Thus, Chabry came to continue the approach of Dareste with the idea of producing monsters experimentally, and that of Warynski and Fol by employing the direct method: he was part of a francophone teratological tradition, while drawing inspiration from the "developmental mechanics" of the German authors (Fischer 1991).

For Chabry, monstrosities, whether natural or produced experimentally, fell into two distinct categories with little in common. In one category, a single *fait initial* is responsible for a causal chain that directs the egg towards producing a malformed or monstrous individual: here experimentation is possible. In the other category, several such initiatory events – for instance death of one blastomere and fusion of two others – are responsible for abnormal development: here experimentation is impossible because the experimenter cannot initiate several teratogenic events at the same time.

In following his teratological investigations, Chabry thought that the spontaneous monstrosities in his ascidian egg batches originated essentially from the state of the egg at fertilization: they depended on hereditary phenomena. To support this tenet he cited certain hereditary anomalies, such as polydactyly in man. The logic of his reasoning led him to distinguish between *monstriparité* and *monstruosité*, the former being of hereditary origin and the latter having a congenital or experimentally induced origin. The experimental production of defects (anomalies, malformations, monstrosities) is possible because the hereditary and the induced death of a blastomere give the same result. Nonetheless, Chabry (1887, p. 88) stressed that *l'identité des effects ne témoigne nullement de l'identité des causes*; in other words, experimental teratology "applied to normal eggs ... remains silent with respect to the primary cause" of anomalies (*la cause première* of course is biological for him, and not of theological inspiration). Other embryologists producing monsters by experimental intervention "did believe thereby to counterfeit nature in every point, although they did nothing but imitate her process more simply" (Chabry 1887, p. 91).

With this restriction Chabry refers also to his own experiments of 1886, in which, aligning his *perforateur*, he had been able to reproduce certain teratological cases that he had observed in his aberrant egg batches. Using his microneedle, he had killed a single blastomere either at the two-cell or at the four-cell stage and – depending on which blastomere was destroyed – he had obtained half-individuals (either left or right, see Figure on the next page) or partial larvae lacking the eye spot or the otolith. By linking such defects to the initial anomaly, he felt able to localize the otolith to the left posterior blastomere of the four-cell stage and the eye spot to the right anterior blastomere. Certainly his experiments demonstrated that organs were predetermined right from the beginning of cleavage divisions. In some way this appeared to confirm those who viewed the egg as a mosaic, notably Wilhelm Roux (1885), for whom, in Chabry's words, "the aim of cleavage is to separate the qualitatively differing contents of the nucleus" and who, according to L. Bard (1886), elaborated a potential *arbre genéalogique* in the fertilized egg. But Chabry, perhaps because of his political creed, cautioned his readers against generalizing the meaning of his findings: "The method of cellular traumatism must not in fact always lead to the same result, and while several persons might see in my experiments a decisive proof for the animal being preformed in the egg, and for each part of the ani-

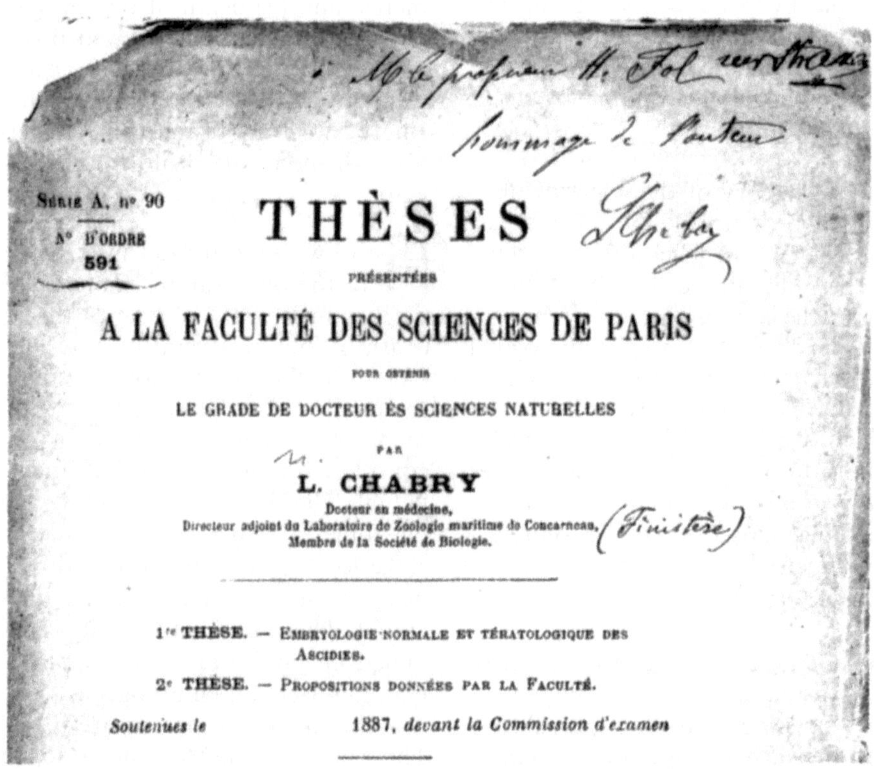
Série A, n° 90
N° D'ORDRE
591

THÈSES

PRÉSENTÉES

A LA FACULTÉ DES SCIENCES DE PARIS

POUR OBTENIR

LE GRADE DE DOCTEUR ÈS SCIENCES NATURELLES

PAR

L. CHABRY

Docteur en médecine,
Directeur adjoint du Laboratoire de Zoologie maritime de Concarneau, (Finistère)
Membre de la Société de Biologie.

1re THÈSE. — Embryologie normale et tératologique des Ascidies.

2e THÈSE. — Propositions données par la Faculté.

Soutenues le 1887, devant la Commission d'examen

mal being preformed in a certain portion of the egg, I tend to reject that rather absolute conclusion" (Chabry 1887, p. 133). He had certainly shown that the ascidian egg was anisotropic, but he refused to consider anisotropy a general rule applicable to the entire animal kingdom. Nonetheless, his findings sparked the debate about isotropy or anisotropy of the egg more intensely than ever, and with it that on neo-preformation and neo-epigenesis.

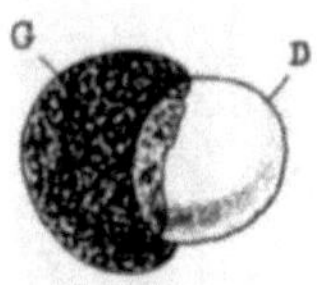

127

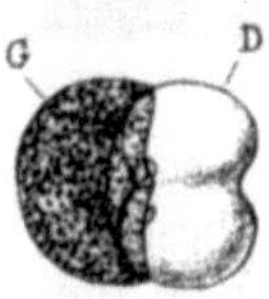

128

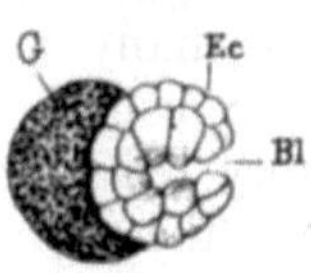

129

Chabry's results were extensively discussed, reproduced and enlarged on by numerous other authors. There was much controversy over the interpretation of his experiments, but this is normal among scientists. Several factors come into play in the interpretation of an experiment: in addition to the results, the theoretical preconceptions and the ideological sensibilities play a not inconsiderable role in biological interpretations. Thus, Oscar Hertwig (1892), a protagonist of neo-epigenesis, interpreted Chabry's results as demonstrating the isotropy of the egg: "In a large number of cases it was possible to obtain, from eggs thus mutilated, completely normal larvae which were lacking nothing but organs of little importance, like an otolith or an adhesive papilla." H. Driesch (1894), at that time still a champion of neo-epigenesis (Sander 1990), was also of this opinion and considered the gastrula shown in Figure 129 in Chabry's thesis (Figure on this page) to be complete and to result from regulation in the egg. On the other hand, Dietrich Barfurth (1893), an adherent of the mosaic egg concept, shared Chabry's opinion, interpreting his Figure 129 as a half-gastrula. Somewhere between these conflicting interpretations lay the theoretical compromise of Wilhelm Roux (1988), who, invoking his theory of postgeneration, considered that Chabry's Figure 129 represented a normal gastrula reduced in size by one-half.

Between Chabry and Albert Dalcq (1932, 1938), who following E.G. Conklin (1905) demonstrated regulative powers in the ascidian egg prior to fertilization, the interpretation progressed towards reconciliation, for it was necessary to accept components of both neo-preformation and neo-epigenesis to explain the events that led to the formation of a living being and its specific continuity. Yet irrespective of their choice of theoretical positions, the pioneers of experimental embryology and their successors must surely have shared a feeling that Chabry (1887, p. 134) expressed thus: "All the experiments, all the mutilations to which one subjects a normal egg contribute ultimately to revealing its structure, and this certainly is one of the finest researches that the naturalist can propose."

All figures are from Chabry (1887): Figs. 127 and 128 show two stages of an embryo after destruction of the left blastomere at the two-cell stage, and Fig. 129 is another half-individual at gastrulation (*Bl* blastopore; *Ec* ectoderm; *D* right, *G* left blastomere). The 1887 reprint (×0.7) dedicated to Hermann Fol is kept in the Central Library of Saarbrücken University; the signature in the top corner is that of Otto zur Strassen, known for his work on *Ascaris* monster embryos. This essay was translated from the French original by K. Sander.

References

Bard L (1886) La spécifité cellulaire et l'histogenèse chez l'embryon. Arch Physiol Norm Pathol [Ser 3] 7:406–420

Barfurth D (1893) Halbbildung oder Ganzbildung von halber Größe. Anat Anz 8:487–493

Chabry L (1887) Embryologie normale et tératologique des ascidies. Felix Alcan, Paris

Conklin EG (1905) Mosaic development in ascidian eggs. J Exp Zool 2:145–223

Dalcq A (1932) Etude des localisations germinales dans l'œuf vierge d'Ascidie par des experiences de mérogonie. Arch Anat Microsc 28:223–333

Dalcq A (1938) Etude micrographique et quantitative de la mérogonie double chez *Ascidiella scabra*. Arch Biol (Liége) 49:397–568

Dareste C (1891) Recherches sur la production artificielle des monstruosités ou essai de Tératogénie expérimentale. Reinwald, Paris

Driesch H (1894) Von der Entwickelung einzelner Ascidienblastomeren. Arch Entwicklungsmech Org [Roux' Arch] 1:398–413

Fischer JL (1987) Camille Dareste (1822–1899). In: Congrès national des sociétés savantes, Histoire des sciences, Lyon. Comité des travaux historiques et scientifiques, Paris, pp 149–155

Fischer JL (1991) Monstres, histoire du corps et ses défauts. Syros-Alternatives, Paris

Fischer JL (1992) Laurent Chabry (1855–1893) and the "Age d'or" of biology. Roux's Arch Dev Biol 201:61–64

Hertwig O (1892) Zelle und Gewebe. Fischer, Jena

Lereboullet A (1864) Recherches sur les monstruosités du Brochet, observées dans l'œuf et sur leur mode de production. Ann Sci Nat [Zool] 1:113–199

Roux W (1885) Beiträge zur Entwickelungsmechanik des Embryo: Einleitung und (1) Zur Orientierung über einige Probleme bei der embryonalen Entwicklung. Z Biol 21:411–524

Roux W (1888) Über die künstliche Hervorbringung halber Embryonen durch Zerstörung einer der beiden ersten Furchungskugeln, sowie über die Nachentwicklung (Postgeneration) der fehlenden Körperhälfte. Virchows Arch 114:133–153; 246–291

Sander K (1990) Von der Keimplasmatheorie zur synergetischen Musterbildung – Einhundert Jahre entwicklungsbiologischer Ideengeschichte. Verh Dtsch Zool Ges 83:133–177

Warynski S, Fol H (1884) Recherches expérimentales sur la cause de quelques monstruosités simples et divers processus embryogéniques. Rec Zool Suisse 1:1–25

ESSAY

Landmarks in Developmental Biology (9)

How to dart ascidian blastomeres: The embryological micro-tools of Laurent Chabry (1855–1893)

Klaus Sander and Jean-Louis Fischer

To realize his plans to eliminate individual blastomeres of the ascidian embryo (see Fischer 1992a, b), around 1885 Laurent Chabry devised a sophisticated set of micro-tools for use under the microscope. It is the aim of this essay to describe these tools, to explain his methods of constructing and employing them and to follow their possible influences on subsequent research and on tools developed by later experimental biologists.

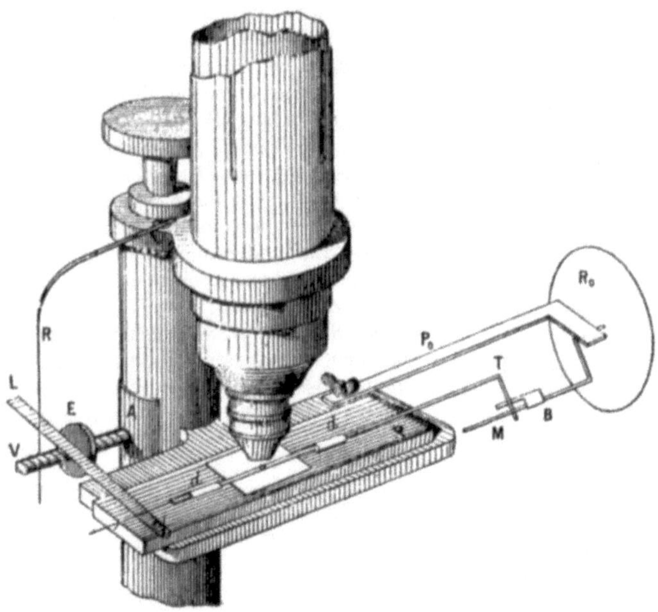

Fig. 1.

Chabry wished to achieve three purposes with his tools:

1. To pick out individual embryos from among a large number of others
2. To orient individual eggs or cleavage stage embryos with a diameter of about 0.16 mm (*Ascidia aspersa*) or even less (e.g. sea urchin eggs) at the experimenter's will
3. To pierce a blastomere through the tough yet elastic egg envelope, without playing havoc with the remainder of the embryo.

He set great store by these tools, as is evident from the very detailed description he gave of them, taking up almost as many pages as he devoted to the results of his experiments (Chabry 1887). Figure 1 (this page) shows the entire assemblage, a striking testimony to his acceptance of the postulate of one his teachers: "The experimenter has to be able, at each moment, to modify his instruments and frequently to construct them himself" (Marey 1878, p. 444).

For picking individual specimens from among large numbers of spawned eggs under the microscope (magnification 50 ×), Chabry constructed a pipet holder permitting the pipet to be slewed horizontally and vertically. By means of a long handle sticking out from the rear end of the pipet, he was able to exercise very fine control over the excursions of the pipet's tip. Suction was generated by a syringe attached to the bench and connected with the pipet by rubber tubing. For fine adjustment and control of water currents, a valve at the syringe's nozzle could be opened so as to curb positive or negative pressure building up in the air-filled tubing. At the outset, the valve was opened to allow water to enter the

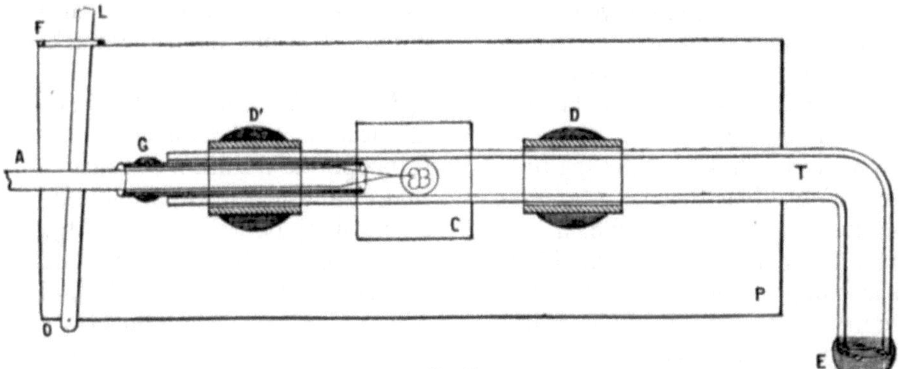

Fig. 2.

pipet by capillary force. At equilibrium, the valve was closed and the egg aspirated by the action of the syringe. When the egg had entered the pipet the valve was opened again, to prevent it from being sucked up too far. After the pipet was slewed out from underneath the front lens, the egg could be ejected onto a slide or into a some vessel positioned nearby. For minute manipulation, the egg could be sucked into a long capillary matching its diameter; for this, the pipet was replaced by a glass tube holding the capillary by a pressure seal inserted in its far end.

In order to take full advantage of this method of mounting the egg, it should be possible to rotate the capillary in addition to shifting its position on the microscope stage. Chabry achieved both ends by using a massive glass slide as a *porte-capillaire* (P in Fig. 2), onto which two bearings cut from rubber tubing were glued (D, D′). In the region harbouring the egg, the *capillaire tourne-objet* (T, not drawn to scale in the figures) was submerged in a drop of distilled water topped by a cover slip to minimize refraction and permit observation at high magnifications (300 ×). One end of the capillary was bent to a right angle, to provide a lever for rotation. When the support was positioned on the microscope stage (Fig. 1), the lever was engaged by a cranked fork (B) that could be rotated coaxially to the capillary by a milled knob (Ro).

The price paid for the ease of handling and orienting the egg in the capillary was the rather limited period of time before development tended to become abnormal (probably as a result of insufficient gas exchange). Under a cover slip, eggs could be maintained (with proper changes of water) for several hours and would continue normal development thereafter, but in a capillary, Chabry noted, they could not be observed for more than about 10 minutes, and this only two or three times a day – one of several troublesome restrictions.

The true technical challenges came, however, when Chabry set about piercing individual blastomeres by a *perforateur*. It was only after many futile attempts that he could write "j'ai résolue complètement ce problème" (1887, p. 10). The solution was based on the capillary holder just described, in combination with some fairly simple but ingenious contrivances attached to the microscope. The *porte-capillaire* was fitted, for this purpose, with a tiny lever (L in Fig. 2) hinged to one edge (at O) and limited in its excursions by two stops at the opposite edge (F).

To obtain *dards* suitable for penetrating the egg covers, Chabry first experimented with minute insect pins and with natural products, such as polychaete setae and sponge spicules. In the end, however, he settled on glass lances. By pressing one end of a fine glass filament onto the heated platinum loop of a thermocauter and rapidly pulling it off as soon as it started sticking to the dark-red glowing metal, he obtained short, stiff needles tapering down to a "point merveilleusement fin" (1887, p. 11). But even after he had systematically studied the effects of temperature and pulling kinetics on the shape and quality of the points, he was to find that for assembling a collection of good lances "il faut s'armer de beaucoup de patience" (Chabry 1887, p. 12).

This commodity was even more in demand when it came to getting the *perforateur* to work. To begin with, the glass *dard* was threaded, blunt end first, into a stretch of capillary serving as a snugly fitting protective barrel (G in Fig. 2). This in turn had been selected to match the inner diameter of the capillary that was to hold the egg. However, the barrel had to be glued in place (Fig. 2) before the capillary holding the egg was pushed through its bearings (D, D′) from the opposite side. Only then was the lance's point, hitherto protected inside the barrel, slid forward to the vicinity of the egg. Lever L was then set to the far position (close to F in Fig. 2), and the glass lance was joined to it by a minute drop of elastic glue.

All this acomplished, the *porte-objet* was transferred to the microscope stage and the lever engaged with a "stepper device" (in present-day terminology) fixed to the microscope stand (Fig. 1). Its purpose was twofold: it had to jerk the lance's point through the resilient egg covers, and it was to stop the lance when the point had passed into or just through the blastomere to be disabled. The jolt was generated by pulling the brass spring, R, a little way off the lever, L, and letting it snap back. The movement thus imparted to the lever came to a stop when the lever hit the nut (E) mounted on a threaded rod (V). At the outset, the nut was positioned so that the lever (positioned at F, see Fig. 2) rested against it. Then, after pulling the spring off the lever, one shifted the nut slightly towards the microscope stand, in which position it would stop the lever jerked forward by the spring. Owing to the lever's scaling effect, the glass lance would make an even smaller but still well-controlled excursion visible under the microscope. Chabry claimed that "avec un peu d'habitude" one could thus perforate an ascidian blastomere to exactly the depth desired (figure marked 6). For piercing even smaller objects, such as sea urchin eggs (see figures marked 135 and 136), he devised an even more intricate version of the *perforateur*.

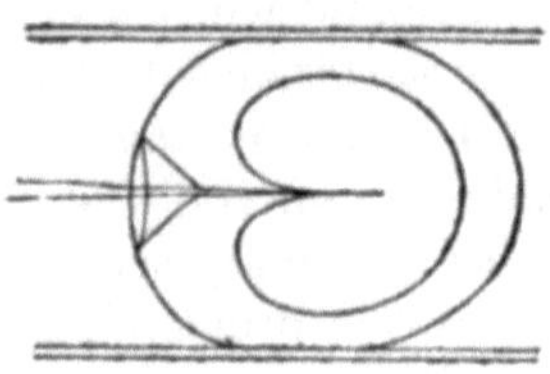

Fig. 6.

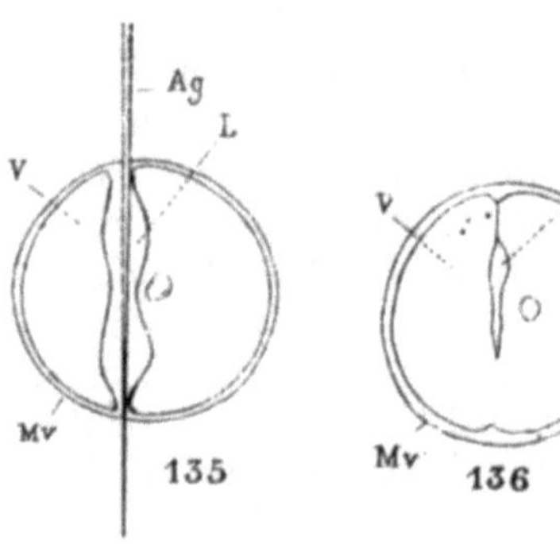

Despite these fine tools, Chabry's conclusions on the mode of ascidian embryogenesis (Fischer 1992b) were based mainly on spontaneous rather than experimentally induced anomalies. Readers who, *avec beaucoup de patience*, have perservered thus far will guess why: the technique was just too demanding. Chabry (1887) said as much on p. 144, in closing the experimental chapter of his thesis: "The considerable time that has to be devoted to each experiment because of the minuteness of the operation and owing to fruitless or lost cultivations [of embryos] did not permit me to produce a number of monsters sufficient to allow a rigorous evaluation of the method, to that degree which one can achieve in research on pure morphology ...".

Therefore, when Edwin Conklin started experimenting on early ascidian development some 20 years later, he chose a different approach: "I have not attempted to repeat the various ingenious methods of injuring blastomeres which were devised and employed by Chabry, since they are necessarily slow and difficult of application and yield but a small number of injured eggs, whereas by simply spurting or shaking the eggs one may injure blastomeres in an enormous number of eggs, which can then be sorted out and classified according to the character of the injury; furthermore, the ease and certainty with which the identity of injured blastomeres of *Cynthia* may always be determined renders unnecessary such experiments as Chabry's on the individual cleavage cells" (Conklin 1905, pp. 154/155). Obviously, by the turn of the century, mass production had replaced the artisan's craftmanship here as in the business world!

Since later students of ascidian embryogenesis employed other techniques, was Chabry's tool-making ingenuity wasted and lost? Certainly not, although subsequent micro-surgery on tiny embryos – notably the sea urchin experiments of Sven Hörstadius (1928) – was performed with glass needles of Spemann's rather than Chabry's make. Whether the technique of mounting *Drosophila* embryos in a capillary (e.g. Schmidt-Ott and Technau 1992) owes a debt to Chabry – who on p. 10 of his thesis envisioned such a technique – may seem more than doubtful, and this holds also for the "touch-and-pull" technique of producing those stiff glass needles that Gerhard Krause (1935) used for microsurgery on insect embryos. A connection seems likely, however, between the *porte-objet capillaire* and a similar contrivance linked to another famous name, that of Horatio S. Greenough, the American who persuaded the firm of Zeiss to construct the first widely used stereomicroscope (Czapski and Gebhardt 1897; Greiner and Sander 1987). For this microscope Zeiss offered a *Kapillar-Rotator*. According to Gause (1963), Greenough wrote his first letter to Ernst Abbé, dated July 4th, 1892, at the *laboratoire de zoologie marine de Concarneau* – the very place where Chabry had devised and used his instruments a few years earlier. Apparently, however, Greenough's gadget also proved impractical, for it soon vanished from the market.

Yet in the long run, Chabry's technical genius lived on in a francophone tradition paralleling that established in experimental embryology (see Fischer 1992b). The art of producing micro-tools was kept alive by Eugene Bataillon, whom Chabry had tought how to make fine glass needles (Fischer 1992a), and it culminated in de Fonbrune's micro-forge (de Fonbrune 1949), which is still in worldwide use today for the production of sophisticated glass tools for micromanipulation.

References

Chabry L (1887) Embryologie normale et tératologique des ascidies. Felix Alcan, Paris

Conklin EG (1905) Mosaic development in ascidian eggs. J Exp Zool 2:145–223

Czapski S, Gebhardt W (1897) Das stereoskopische Mikroskop nach Greenough und seine Nebenapparate. Z Wiss Mikroskopie 14:289–303

de Fonbrune P (1949) Technique de micromanipulation. Masson, Paris

Fischer JL (1992a) Laurent Chabry (1855–1893) and the "Age d'or" of biology. Roux's Arch Dev Biol 201:61–64

Fischer JL (1992b) The embryological oeuvre of Laurent Chabry. Roux's Arch Dev Biol 201:125–127

Gause H (1963) Siegfried Czapski 1861–1907. In: Freund H, Berg A (eds) Geschichte der Mikroskopie, vol III. Umschau Verlag, Frankfurt/Main

Greiner K, Sander K (1987) Das Stereomikroskop – Ursprünge und geschichtliche Entwicklung. Biologie in unserer Zeit 17:161–168

Hörstadius S (1928) Über die Determination des Keimes bei Echinodermen. Alb. Bonniers Boktryckeri, Stockholm

Krause G (1935) Analyse erster Differenzierungsprozesse im Keim der Gewächshausheuschrecke durch künstlich erzeugte Zwillings-, Doppel- und Mehrfachbildungen. Roux' Arch Entwickelungsmech Org 132:115–205

Schmidt-Ott U, Technau GM (1992) Expression of *en* and *wg* in the embryonic head and brain of *Drosophila* indicates a refolded band of seven segment remnants. Development 116:111–125

The figures are from Chabry (1887), with Fig. 135 showing a pierced sea urchin egg that survived after withdrawal of the *dard* (Fig. 136) and finally yielded a complete larva.

ESSAY

Landmarks in Developmental Biology (10)

Shaking a concept: Hans Driesch and the varied fates of sea urchin blastomeres

Klaus Sander

Among the early adherents of *Entwickelungsmechanik,* Hans Driesch (1867–1941) was certainly the most colourful personality. Born at Kreuznach, a well-known spa on the Nahe river, as the only child of a wealthy Hamburg merchant, he grew up in the liberal atmosphere of that hanseatic city (Driesch 1951). Gifted with the self-confidence that results from a strong mind backed by seemingly unlimited financial means, and driven by a brillant intellect formally trained at the best institutions within reach, he had by his mid-twenties published nearly two dozen descriptive and experimental papers on animal development, and three small books in the border-fields between theoretical biology and philosophy of science. Many of these have fallen justly (and some unjustly) into oblivion, while two or three proved seminal. These last dealt with the shaking and compressing of cleavage stage sea urchin eggs, and served more than any others to shake confidence in the then widely accepted concept of mosaic development (see the previous essays in this series).

A cosmopolitan at heart and in fact, living alternately in Zürich in Switzerland and on the shores of the Mediterranean, Driesch by that time had seen a good deal of Europe, travelled in Asia Minor and Egypt, and spent several months in India, Ceylon and Java on two visits. On these travels, Driesch had collected hydrozoans from many coasts. He soon became interested in the "laws of growth" that shaped their colonies, and he chose to use this material for a doctoral thesis that he prepared rather independently in Ernst Haeckel's laboratory at Jena.

In a kind of summary of this work, Driesch (1891 a, p. 21) is seen to look forward to his next stance: "If for just a single animal its origin from cells could in this way be grasped and represented in [mathematical] formulas, as done here for the construction of colonies from their units [the individuals], then our knowledge of organismic forms would at least be on the pathway permitting one to hope for future insight". He collected all available data on cleavage patterns – only to find that what he had foreseen was really true, namely "that the problem of a geometrical animal ontogeny is a much more complicated one than that with which we deal here ..." (p. 21). He gave up – and then by chance happened across Roux's "Die Entwickelungsmechanik der Organismen, eine anatomische

Fig. 2.

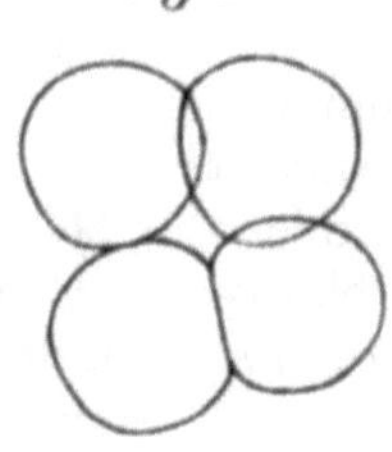

Fig. 3.

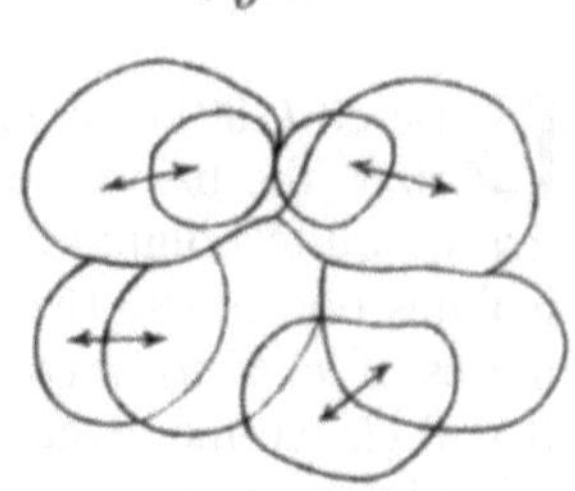

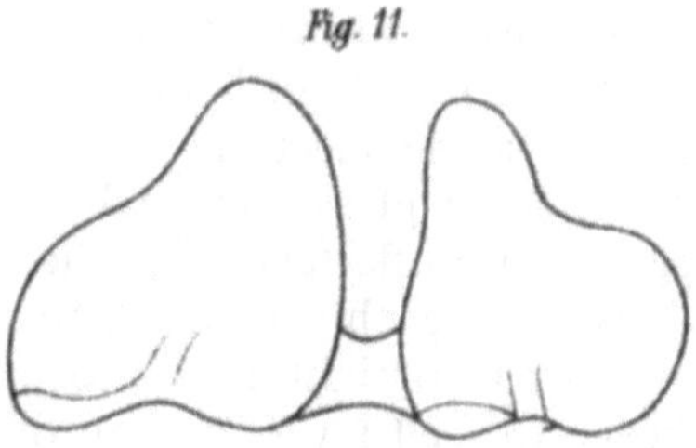

Fig. 11.

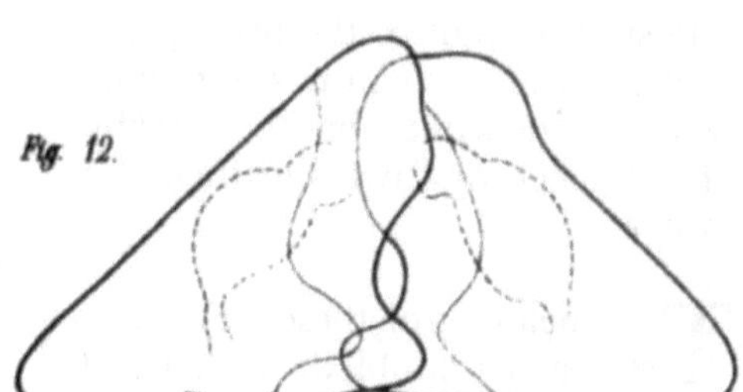

Fig. 12.

Wissenschaft der Zukunft" (see Essay 1). Despite some contempt for the name of the new discipline, he decided to analyze sea urchin development.

From earlier authors, Driesch knew that by shaking the unfertilized sea urchin egg one might obtain fragments of the egg cell. He decided to go to Trieste on the Adriatic and find out whether shaking could also serve to separate the blastomeres of the two-cell stage. Driesch chose this stage because Fol and the Hertwigs had shown that polyspermy causes the egg cell to divide simultaneously into more than two blastomeres. Using two-cell stages, he could feel assured that only a single sperm had fertilized the egg – a circumstance worth considering because of the then current view that twinning in mammals might result from dispermic fertilization.

Shaking many batches of about a hundred eggs vigorously in small vials for several minutes, in the course of several weeks Driesch amassed about 50 fully isolated and viable blastomeres. These he raised individually in small dishes.

On reading Driesch's accounts (1891 b) of sea urchin cleavage, whether in separated blastomeres or in whole eggs, the modern reader will develop the uneasy feeling that something has gone awry – a feeling frequently met with on reading old literature. In the present case, as with Roux's split embryos (see Essay 4), this feeling stems from a faulty fate map. Following the earlier descriptions of Selenka (see figure of the 16-cell stage on the front page), Driesch considered the micromeres to mark the animal pole while the archenteron, an acknowledged vegetal structure, was to develop from the mesomeres at the opposite pole. It took him another eight years and a host of confusing results to realize that the micromeres form vegetally, i.e. at the very pole that gives rise to the archenteron.

Cleavage of isolated two-cell stage blastomeres in most cases was partial, each of the resultant cell tiers consisting of only half the normal cell number, as shown in Figs. 2 and 3 for the half-8 and the half-16 cell stage. Following a half-32 cell stage, the embryo usually developed into an open half-blastula (Fig. 5 below). Much to Driesch's surprise, however, this would close afterwards and give rise to a morphologically normal dwarf-gastrula (15 cases) or even dwarf-pluteus (3 perfect cases). Thus, development until the blastula stage corresponded to "mosaic work" similar to that claimed by Roux for his frog hemi-embryos (see Essay 5). But then a normal course of development would take over – and this not by postgeneration, but evidently by changing cell fates (see below).

Fig. 5

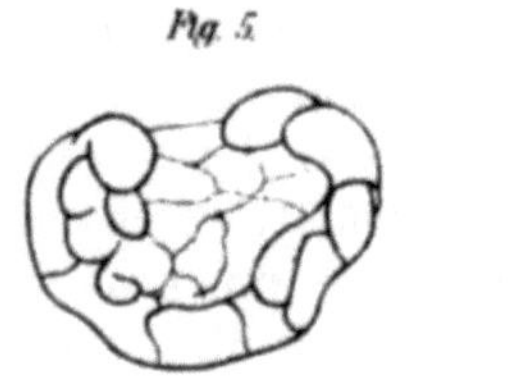

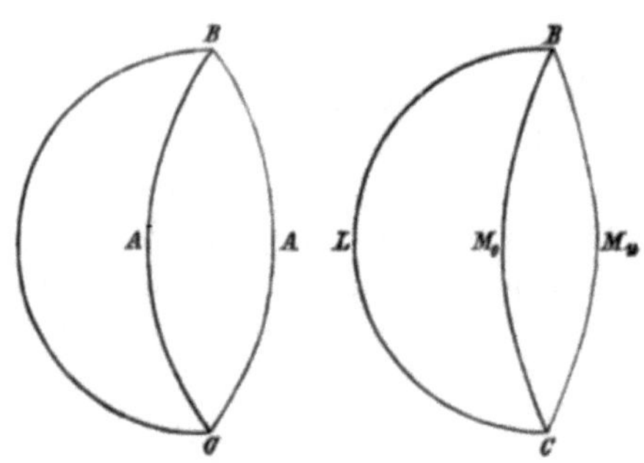

Fig. I. Fig. II.

Driesch (1891 b, p 178) summarized the harvest of his stay in Trieste quite poignantly: "If one isolates one of the two first blastomeres of *Echinus microtuberculatus*, it will cleave like a half-embryo [Halbbildung] but yield a complete individual of half size ... Thus for the species studied, the principle of organ-forming territories is refuted while at the same time the possibility of generating twins artificially is proven."

The meaning of these statements cannot be doubted, but part of the message can. For neither confirmed the results that (as claimed expressly in the paper, p. 173) *each* blastomere of each pair can yield a complete individual: the only proven case of twinning was the pair of Siamese twins shown in Figs. 11 and 12. Nor was the principle of organ-forming territories refuted, because as defined by Wilhelm His and understood by Roux (see Essay 1), that principle principle concerns the descent of organs in normal development, not the timing of specification and determination of the territories that give rise to those organs.

During the next couple of months, Driesch was to extend and modify his approach, both technical and conceptual. Experiments in Naples in the autumn of 1891 and spring of 1892 gave much better yields of isolated blastomeres and of survival. He published his findings in three interlinked papers ("Entwicklungsmechanische Studien III–V"), followed by a general discussion of the theoretical aspects (VI).

The theoretically most important outcome of the various new experiments was the fact that quite devious courses of cleavage can lead to normal subsequent development. Driesch obtained nice plutei from eggs whose blastomeres had become deranged after heat shock (about 12 h at 31° C) or in embryos flattened by compression.

The compression experiments (IV) were initiated in order to test O. Hertwig's claim that in flattened eggs the spindles will always orient at right angles to the smaller diameter. Driesch realized that this must lead to a dramatically altered lineage of cells or at least their nuclei. Once 8, 16 or 32 cells had formed in a single sheet, he removed the cover glass under which the egg had been flattened. As the next division in each cell would invariantly occur at right angles to this sheet, two cell layers on top of each other would result. As shown on this page, the 4-cell stage (β 1) under compression will convert into a flat 8-cell stage (2). After release from pressure, this will become double-layered during the next division (3). Of the 16 cells, usually only 1 or 2 would be of micromere size, shown here in faint outline as A and B, covered partly by the bold sister cells, A and B, of the upper layer. Driesch's crucial argument is best grasped from Figs. C and D, showing the rare type with 4 micromeres. The mesomeres (labeled E, F, G, H in Fig. 3) are seen to occupy the opposite poles of the elongated cell aggregate, being arranged there in square (C) or linear configurations (D). Since Driesch, as explained above, believed the archenteron to derive from the mesomeres, he expected to see a somewhat cigar-shaped gastrula with an archenteron at each end on the next day. What he got, however, was a single archenteron and a uniform and normal-looking pluteus.

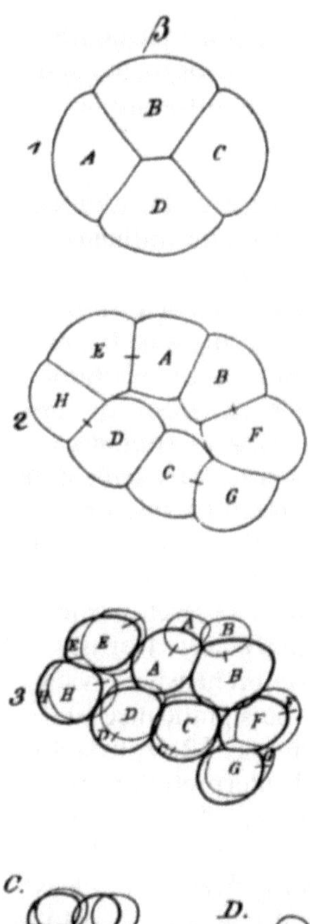

From this and similar results, Driesch drew the sweeping conclusion "that the cleavage globules of the echinids are to be considered an equivalent material which one may reshuffle haphazardly without causing normal development to suffer in the least" (Driesch 1892, p. 27).

Compare this claim to his thoughtful discussion, in the 1891 b paper, of the half-blastula diagrams shown on the previous page. Figure I allowed him to demonstrate that the closing of the half-blastula could formally occur in two ways, either by point B approaching point C, or by points A moving towards each other. Driesch decidedly preferred the latter mode, because it would preserve each of the original tiers of blastomeres (albeit with only half the cell number). "One can immediately grasp the importance of this question if one considers the difference that the respective concentric rings [the tiers] reveal at least during the early course of cleavage in echinids" (Driesch 1891 b, p. 169). He also stressed that with this mode, but not with the other, the original egg poles would be maintained. Figure II was intended to demonstrate that in any case the closing of the half blastula must involve a change in cell fates. For if M_o and M_u, two points on the future midline, were to meet when the gap closed, they would find themselves forming the right hand edge of the developing larva (L is intended to mark the original left margin, thought to maintain this function in the half-blastula). Conversely, if C were to move upwards and meet B, the original left flank (L) would wind up forming the lower pole.

"Whichever twist one gives to this matter, one cannot get around the profound difference in the roles that a given part of the germ has to play depending on whether a single individual arises or two..." (Driesch 1891 b, p. 172). Thus the confidence in mosaic development was shaken in its very foundations. Yet it was only in the theoretical paper (VI, p. 39) of the 1892 series that Driesch drew the conclusion that is considered his hallmark to this day: "The relative position of a blastomere within the whole will probably in a general way determine what shall come from it; if it be situated differently, then it will give rise to something else; or stated in another way: its prospective relation is a function of [its] place".

This conclusion, in the words of E.B. Wilson (1893, p. 607) "rests upon a series of experiments so ingenious in conception, and brilliant in results, as to stand beside those of Roux as the mark of a new era in the history of embryology".

References

Driesch H (1891 a) Die Stockbildung bei den Hydroidpolypen und ihre theoretische Bedeutung. Biol Zentralbl 11:14–21

Driesch H (1891 b) Entwicklungsmechanische Studien. I. Der Werth der beiden ersten Furchungszellen in der Echinodermenentwicklung. Z wiss Zool 53:160–178 (issued November 20th 1891, volume completed in 1892)

Driesch H (1892) Entwicklungsmechanische Studien III–VI. Z wiss Zool 55:1–62

Driesch H (1893) Zur Verlagerung der Blastomeren des Echinideneies. Anat Anz 8:348–357

Driesch H (1900) Die isolirten Blastomeren des Echinidenkeimes. Arch Entwicklungsmech Org [Roux' Arch] 10:361–410

Driesch H (1951) Lebenserinnerungen. Aufzeichnungen eines Forschers und Denkers in entscheidender Zeit. E. Reinhardt Verlag Munich Basle

Wilson EB (1893) Amphioxus, and the mosaic theory of development. J Morphol 8:579–617

The portrait shows Driesch in 1902. The figures on the first and second pages are from Driesch 1891 b, and the third figure is from Driesch 1893, but similar figures were published by him in 1892. All translations by the author.

ESSAY

Landmarks in Developmental Biology (11)

Hans Driesch the critical mechanist: "Analytische Theorie der organischen Entwicklung"

Klaus Sander

In the realm of biology, Hans Driesch (1867–1941) is known for his early experiments on sea urchin embryos (see the previous Essay), and for his later credo that life – and particularly ontogenesis – depends on some immaterial vital force and thus cannot be fully understood in terms of physics and chemistry; that indeed it defies any materialist approach (the topic of the next Essay). What is rarely mentioned are his early theoretical writings on embryogenesis, only traces of which – such catchwords as *prospective value* and *prospective potency,* and the *Fundamentalsatz* that a cell's fate is a function of its position – have survived, the last owing to its happy re-incarnation in the mind of Lewis Wolpert (Wolpert 1986).

The widespread neglect of this aspect of Driesch's work has been noted before, albeit only by those who ponder on the past writings of biologists as much as on their present plights: "Why do we now place so little emphasis on Roux's heavy emphasis on the interrelationship of parts, and on Driesch's remarkable materialistic analysis? Perhaps in part because we no longer read what they wrote, but in part also because these were early, not late concepts in the minds of Roux and Driesch themselves... But [Driesch's] early words too may have had their influence on the great minds that must have been exposed to them in the happy days when there seems to us to have been more time for reading and contemplation" (Oppenheimer 1965).

The present essay is devoted to the *Analytische Theorie der organischen Entwicklung,* the key work of Driesch the critical mechanist, a booklet of about 180 pages (Driesch 1894). Driesch's aim is to break down, by "causal analysis" (one of Roux's terms), the continuous process of animal morphogenesis into its ultimate elements (or first principles, one might say). However, his critical mind, well trained in philosophy, soon led him to postulate a grave epistemological restriction on this enterprise, which he discusses at great length in the book (see the next Essay). Here I shall concentrate on some of Driesch's mechanistical considerations and try to mark the visionary highlights of his "theory of epigenetic evolution" (a term explained in Essay 6). In a nutshell, Driesch characterized this theory as follows: "Development starts with a few ordered manifoldnesses; but the manifoldnesses create, by interactions, new manifoldnesses, and these are able, by acting back upon the original ones, to provoke new differences, and so on. With each response, a new cause is immediately provided, and a new specific reactivity for further specific responses. We derive a complex structure from a simple one given in the egg. The nucleus, although carrier of all [developmental] possibilities, need not embody these in a defined structure; consequently our theory is epigenetic with respect to structure as such" (p. 86).

In his foreword (p. VIII), Driesch stresses the profound influence that Curt Herbst (1866–1946) had on his thoughts. The friendship with Herbst was a decisive factor in the shaping not only of this book but of Driesch's life, from the time when they met first as students at Jena until long after Driesch's marriage in 1899. Herbst's role as Driesch's *alter ego* is evident in his obituary for Driesch (Herbst 1942). At the time when Driesch wrote the *Analytische Theorie,* Herbst published a comprehensive review, entitled "On the significance of stimulus physiology [Reizphysiologie] for causal interpretation of processes in animal development" (see Oppenheimer 1991). In this he drew a parallel between the release of physiological reactions in plants and animals by various stimuli (wellknown by then) and the release or induction of ontogenetic steps of morphogenesis.

Driesch assigns this concept a central role in animal development and defines it in quantitative and qualitative terms. The decisive *quantitative* criterion for released events is the difference between the amount of energy introduced by the releasing stimulus and the amount liberated in the released effect – the matchstick and the cask of gunpowder provide an example implicitly adduced in the book. The *qualitative* criterion might nowadays be said to rely on the informational content of the reacting system; in Driesch's words (p. 49), the "nature of the reacting body" is always (at least partly) responsible for the type of reaction whenever qualitative aspects matter. A property common to both criteria is that (at least in Driesch's time) the reaction could not be predicted from first principles but only from experience. Applied to ontogenesis (p. 49), this means 'that the specific cause for any ontogenetic change is but an impetus that lets pop up, so to speak, an effect which is both qualitatively and quantitatively different; and not only the "quale" of the effect rests within the reacting component, but also the energy involved...'.

The quality of the reaction, which in Driesch's view is an exclusively chemical one, is due to the nucleus. The nucleus remains totipotent throughout life. "The facts of regeneration... and other things enforce the assumption that the possibility [=potential] of forming the entire organism is present not only in each cleavage cell – as proven – but equally in each cell of the body at large, and is only veiled, so to speak, by certain conditions. This potential in our opinion rests with the nucleus, and consequently the nucleus in each cell must maintain its totality" (pp. 47/48). Therefore, in directing the cell's developmental fate, the nucleus "has to be put into action in such a way that it nonetheless maintains its totality" (p. 48). Driesch (p. 88) provides for this by assuming "that the organogenetic ... chemicals *do not originate* directly from the nucleus but arise *under its guidance* in the cytoplasm. *This guidance we consider a fermentative [=enzymatic] effect.* We view the nucleus as a *mixture* of ferment-like compounds, each of which represents a single type of elementary process in that particular ontogenesis. A given releasing event will prompt a certain ferment into action, and this will be a specific ferment, since the protoplasm, the intermediary between stimulus and response, has a specific chemical constitution."

That cytoplasmic constitution, the result of a sequence of previously released processes, defines the cell's *prospektive Potenz* (or its possible fates), because it decides to which future developmental stimuli the cell can react and to which it cannot. To put this in modern terms, the chemical state of the cytoplasm defines the cell's competence.

The notion that the cytoplasm acts as a selective intermediary between (cell-)external stimuli and the nucleus, and how, is one of Driesch's crucial speculations. On pp. 89/90 he explains his concept with a sketch (marked Fig. 8) and a table.

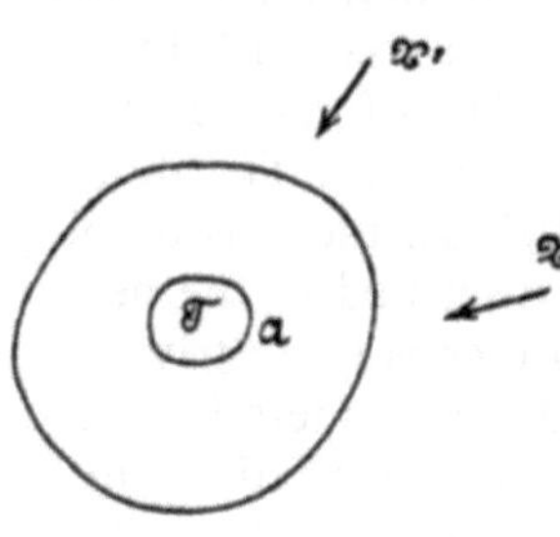

"The true and real complexity of our theory of the releasing process, as opposed to a naive view, as it were, rests in the assumption that the station for the reception of the original organogenic releasing stimulus is topographically separated from the station for the response to it. ... I may be permitted to summarize my concept in a diagram. *A* [*a* in the sketch] is the protoplasm, *T* the totality of elementary potencies, i.e. the nucleus, X_1 and X_2 are sources of inducing stimuli."

The "naive" concept (see the table) treats the cell as a black box, while Driesch views the cell as a system of (at least) two components: in order for the cell to be receptive for a new stimulus (X_2), the previous stimulus (X_1) has to change the cytoplasm so that it partially activates the nucleus, and it is the resulting specific nuclear activity that makes the cytoplasm receptive for stimulus X_2. Since this conceptual framework of a hundred years ago is purely formal, we might today be tempted to endow state A′ of the cytoplasm with a second messenger and to imagine that A_1 differs from A in the presence of some receptor expressed in consequence of the induced differential gene activity *Ta* – a frightening anachronism that might, however, be excused if it helped the modern reader to understand Driesch's concept and, if so disposed, to marvel at his powers of reasoning and his inspiration.

The sequence of events now is this:

The events	Preconditions for the events to be possible
X_1 converts [the state] A into A'	A is receptive for X_1
A' activates T partially to [the state] Ta	T is receptive for A'
Ta converts [state] A' to A_1	A' is receptive for Ta
X_2 converts A_1 to A'_1	A_1 is receptive for X_2
and so on	and so on

In a naive, insufficiently analyzed view our diagram would read as follows:

X_1 converts [the cellular state] Z to [the state] Z_1	Z is receptive for X_1
X_2 converts Z_1 to Z_2	Z_1 is receptive for X_2
and so on	and so on"

To complete this guided tour through the Analytical Theory, we shall sketch Driesch's speculations on the role of induction – he uses this term here in preference to "release" – in multicellular pattern formation. Referring to his *Fundamentalsatz* of 1892, he complains that his critics had unjustly accused him of entrusting position as such with the task of directing the cells' fate. He now pleads – convincingly, in view of his evident mathematical inclinations – that the word "function" was meant in a purely mathematical sense ($x=f[y]$). "We need to resolve position into induced effects, because only a really induced state, one initiated by a [chemical] agent, is a causally effected one" (pp. 53/54). In the gastrula (marked Fig. 7), a cell in the future ectoderm or body wall can be identified by its "Abstandsmaasse x and y" (p. 52) from one of the poles and from some point on the circumference, say where it is cut by the median plane; what Driesch invokes here is obviously a polar coordinate system (he uses the term *Koordinatensystem* in 1898). The cell's fate, however, will not be determined by the coordinates but by inductive stimuli emanating from certain other locations defined by their coordinates in the system. Since the (assumed) inductive stimuli emitted by the archenteron tip will presumably differ from those sent out by the epidermis, a cell, if located at x, will be exposed to a different combination of stimuli than if located at y. "The cell was capable of reacting to both inductions, but it followed that stimulus which owing to its position was the stronger, or might have been the only effective stimulus" (pp. 83/84).

While Driesch here is envisioning inductive events within a continuous epithelium, he also foresaw inductions from one epithelium to another. An example is given in Fig. 5. On imagining a larva

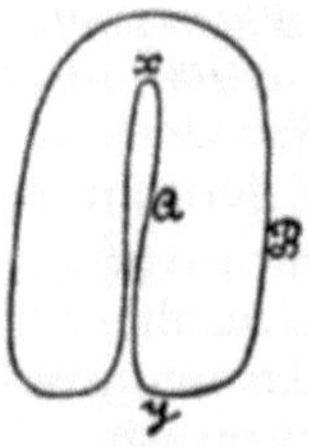

Fig. 7.

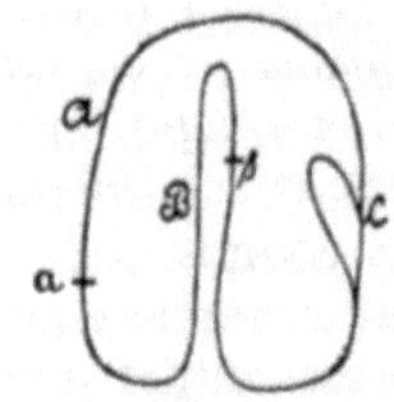

Fig. 5.

formed of the three organs [or anlagen] A, B and C, and remembering that A, B and C differ chemically, "the possibility must be admitted that the C chemical acts on organ B so that where it first reaches B, for instance in the course of diffusion, it releases in B an organ formation for which B was predisposed by its prospective potency. This [induction] would then occur next to C, at β. Note that on this assumption the location of β would change with the position of C ..." (p. 58). What a remarkable vision of discoveries to come in subsequent decades!

These examples should suffice to illustrate Driesch's mechanistic interpretation of ontogenesis in the *Analytische Theorie*. Why did this inspired book fail to acquire the wide recognition that for instance Wilhelm Roux earned with his (equally speculative) manifestos? I think for two reasons. The more profound of these, the epistomological, will be discussed in the next Essay. The other is quite trivial: Driesch made his readers feel very clearly that he considered his intellect superior to theirs. 'Almost nobody has taken notice of my theoretical writings; may this always remain so. This book, too, is not "interesting" and "modern". I do not believe that it is easy to read; indeed I hope that a superficial reader will not at all understand it' (p. VII). His wish was apparently granted, for two years later (Driesch 1896, p. 365) he wrote that a certain error in the book "has led nobody astray, if only because no one has read my book with some care." The one exception he mentions was Taddeusz Garbowski (1895) who in closing his review of Driesch's work (p. 332) wrote: "Future biologists will certainly smile at the attempt to solve ontogenetical riddles at a time when one was equipped only with a metaphysical array of undefined anlagen, potencies, materials, and ferments. Yet the perseverance with which this earnest researcher has carried his mental toils, starting from theoretical considerations and passing through experimental investigations into the realms of the most rigid abstraction, enforces respect."

The quotations (unless indicated otherwise), the figures and the table (translated) are from Driesch (1894). Square brackets mark additions inserted by the present writer.

References

Driesch H (1894) Analytische Theorie der organischen Entwicklung. Engelmann, Leipzig

Driesch H (1896) Die Maschinentheorie des Lebens. Biol Centralbl 16:353–368

Garbowski T (1895) Kausalanalytische Theorie der epigenetischen Evolution mit dreifacher Rhythmusharmonie in der Ontogenese. Biol Centralbl 15:305–332

Herbst C (1942) Hans Driesch als experimenteller und theoretischer Biologe. Roux's Arch Dev Biol 141:111–153

Oppenheimer JM (1965) Questions posed by classical descriptive and experimental embryology. In: Moore JA (ed) Ideas in modern biology. Natural History Press, Garden City, NY, pp. 205–227

Oppenheimer JM (1991) Curt Herbst's contributions to the concept of embryonic induction. In: Browder L (ed) Developmental biology: a comprehensive synthesis, vol 7: Gilbert S (ed) A conceptual history of modern embryology. Plenum, New York

Wolpert L (1986) Gradients, position and pattern: a history. In: Horder TJ, et al. (eds) A history of embryology. Cambridge University Press, Cambridge, pp 347–362

ESSAY

Landmarks in Developmental Biology (12)

Hans Driesch's "philosophy really ab ovo", or why to be a vitalist

Klaus Sander

Had Hans Driesch died while his *Analytische Theorie der organischen Entwicklung* (see the previous Essay) was in the press, he might have become the Franz Schubert of developmental biology: a genius and *frühvollendet*, with unheard-of treasures sprung from his creative imagination waiting to be acknowledged and admired by posterity. Instead, he lived on happily for nearly half a century. Yet the fates asked a price for this: he was to ruin his reputation – that is, among those professing the mechanistical creed – by preaching vitalism.

Not that Driesch lived in misery thereafter – not at all. His material existence was safeguarded by his inherited fortunes, and whatever reputation he may have lost among righteous biologists was compensated many times over by the increasing interest that the educated public took in his philosophical and later on in his parapsychological thinking and writing (Driesch 1951, Oppenheimer 1971).

In 1900, he gave up his nomadic lifestyle and settled in Heidelberg where he raised his children (who both turned to music as a profession). It was only in 1909, shortly before he published his last experimental paper, that he sought an affiliation to the university. Entirely on his own terms – a mark of the reputation he enjoyed by then as a philosopher – he quickly achieved his *Habilitation,* permitting him to teach Natural Philosophy in the faculties of both science and philosophy. Curt Herbst, his lifelong friend, had earlier obtained a readership (later a chair) of zoology at the same university, which he held until retirement, braving the Nazi establishment during his last active years. Driesch, the convinced cosmopolitan and pacifist, also had a strong dislike for the new ideology, and it was fortunate that he obtained tenured appointments before Hitler took over. After a brief spell as a professor of philosophy in Cologne he was offered the chair of philosophy in Leipzig, one of the most renowned in Germany. These appointments were also welcome for financial reasons, because World War I and the subsequent runaway inflation had destroyed the family's wealth. Driesch kept his chair in Leipzig until autumn 1933, when the new rulers forced him to retire.

Before moving to Leipzig in 1923, Driesch had taken a year off to lecture by invitation of the Chinese government at various Chinese unversities, returning to Germany via Japan and the USA where he visited his friends of bygone Naples days, T.H. Morgan, Ross Harrison and E.G. Conklin, and gave lectures in many places. This was his most extensive tour, but many paid invitations to speak on philosophical problems were to follow, among them another one to the USA where this time he met E.B. Wilson and F.R. Lillie in addition to the old friends already mentioned. He lectured in South America and in nearly all European countries, and served twice as the Honorary President of the International Congress of Parapsychology (known in the vernacular as occultism, a designation Driesch vehemently rejected). Apparently, with his cosmopolitan outlook and the formidable list of languages that he mastered, Driesch was welcomed internationally as a representative of Germany's best cultural traditions – but even without this bonus, turning vitalist apparently paid off in those days.

In 1927 Hans Spemann, Driesch's junior by two years, greeted the hexagenarian in his *Festschrift* (two volumes of this Journal) as the "philosophically inclined biologist out of whom the biologically educated philosopher developed". And when conferring a honorary doctorate on Driesch in 1910, the Dean of the Law Faculty at Aberdeen alluded to this metamorphosis with the phrase quoted in the title of this Essay. He must have known, because the academics of Aberdeen had had the benefit, two years earlier, of a course of twenty lectures that Driesch gave on "The Science and Philosophy of the Organism", under an endowment from the late Lord Adam Gifford for a chair of "Natural Theology". In these lectures, published both in English (1908) and in German (*Philosophie des Organischen*, 1909), he propagated his concept of life, known as Neo-Vitalism (*neo-* as distinct from the *bona fide* vitalism holding sway before the rise of cell and developmental biology late in the nineteenth century). As implicated in the Dean's aperçu, Driesch argued his vitalism from the angle of his experiments on sea urchin eggs – or rather from his interpretation of their results. This distinction is important, because his results were open to other interpretations (see the next Essay) and thus did not constitute the compelling evidence for the "autonomy of life" that Driesch saw in them.

Fig. 1.

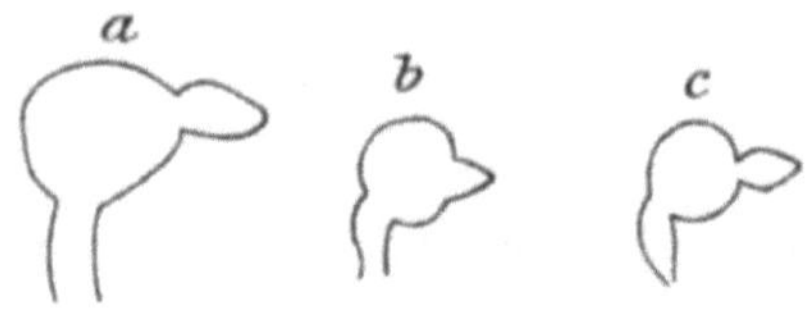

Gegliederte Därme junger Plutei von Sphaerechinus granularis (kopirt aus meiner Arbeit »Zur Analysis der Potenzen embryonaler Organzellen«).
a normales Objekt. *b* und *c* Objekte, welche auf dem vollendeten Gastrulastadium annähernd im »Äquator« durchschnitten waren, so dass nur die Hälfte des Entoderms (und des Ektoderms) belassen ward. Die Differenzirung ging trotzdem im Wesentlichen in typischer Proportionalität vor sich.

We shall now try to trace the main arguments that successively lead Driesch towards being a vitalist, ultimately of the metaphysical brand. His conversion, like much of his other reasoning, was not a straightforward or one-step process, and it was not free of inconsistencies. In his earliest sea urchin papers he mentioned only briefly that development must be subject to specific laws. In the Analytical Theory he became more outspoken, postulating an autonomous status for biology among the natural sciences, based on "teleology" (the apparent purposiveness of structure and function), a property we recognize only in living beings. In the Introduction he writes: 'To begin with, I was attracted by those few experiments which call themselves exact; from there I was led on to an analysis of these "mechanical" intentions, but also to doubts concerning their value for the really biological questions. ... And on continuing this course of thinking it could not escape my notion that exact biology, i.e. predominantly the exact study of [organismic] form, must be placed not as a new discipline within physics but rather as an "independent basic science" at the side of all physics, of all mechanism. ... I recognized the justification and importance of teleology and its relation to the concept of causality; old truths, but nearly strangled by modern superficiality ...'. This complaint refers to Immanuel Kant's *Kritik der Urteilskraft* (1790). Kant held that the "mechanism of Nature" is insufficient for generation of an organized being and therefore it must be subordinated, at least in our categories of recognition, to some intentionally acting cause; yet the mere teleogical cause in turn is insufficient for such a being to be considered a natural product. "The possibility of this fusion of two entirely different kinds of causality ... is beyond our understanding"; in Kant's (and Driesch's) view it rests on the fact that our grasp of reality is severely restricted by our a priori categories of possible understanding. But whereas Kant considered ontogenesis *as if* a purposeful intellect were at work (the teleomechanist's view, see Lenoir 1982), Driesch dropped the "as if" when turning vitalist.

For himself, Driesch discovered teleology via the insight that ontogenesis must be viewed as a cascade of released reactions (see Essay 11). He clearly recognized a concomitant problem of formidable dimensions: how did it come about that those released reactions of cells or tissues were purposeful, i.e. were functionally suitable or even optimal for construction of a viable organism? Or, in the words of Edwin Conklin (1929, p. 31), "how can one explain the apparent teleology of development where the end seems to be in view from the beginning?" Driesch's peers, captive as they were in the "mechanist dogma", did not even see the problem: "... they think mostly of the [proximate] causes of an event but fail thereby to notice that [component] for which no origins can be named" (Driesch 1894, p. 119). Origins can of course be named (at least by way of speculation) for "purposeful" reactions in ontogenesis – provided one accepts the modern Darwinian mode of evolution, with mutations generating variants among which selection favours the (in their context) most "suitable". But mutations as the source of variation were not envisioned at that time, so Driesch (as some other biologists at the turn of century) had easy play in exposing Darwin and his followers to ridicule.

Vitalism came up some years later. In a treatise on "The Machine Theory of Life" (1896), Driesch pretended to defend himself against charges of vitalism: 'Thus, what I proposed (in the Analytical Theory) was not at all "vitalism", but was ... exactly the *common opinion of physico-chemical dogmatism!* Only, I did not hesitate to see and pronounce the consequences of this dogmatism ...: the *giveness of the purposeful basis* on which the events of life come to pass' (Driesch 1896, p. 365, his emphases). And (p. 364): 'On the basis of these given properties, of this "machine", we understand, with the aid of physics and chemistry, its functioning causally very well, the working of the developmental not less than that of the physiological "machine". A few pages later (p. 367), however, he drops the disguise: 'In [stating] this I do not intend to defend my [earlier] opinion – the machine theory of life – as

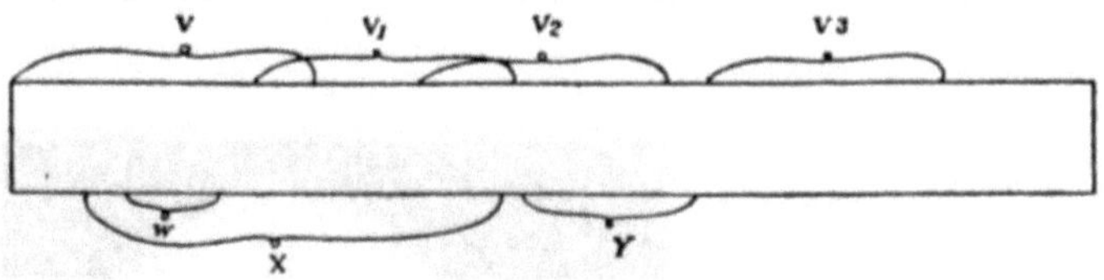

FIG. 14.—AN "HARMONIOUS-EQUIPOTENTIAL SYSTEM" OF WHATEVER KIND.

According to the "machine-theory" of life this system ought to possess a certain unknown very complicated machine *in its completeness:*
(a) in its total length,
and (b) in each of the equal volumes v, v_1, v_2, v_3 and so on,
and (c) in each of the unequal volumes w, x, y, and so on,
and (d) in every imaginable volume, no matter of what size.
Therefore the "machine-theory" of life is absurd.

being correct; I only wanted to make clear what I did say [in 1894]; that alone is the purpose of this treatise. ... I just wanted to say once what one really means if one denies specific laws for the process of life ... Could not the unsatisfactory aspects of this view, which leaves us with so much [teleological] "giveness" ... be eliminated by a really "vital" theory?'

That theory Driesch (1899) developed starting from the problem of localization, the question as to why certain steps in development occur at defined locations – defined, that is, with reference to the whole (according to his *Fundamentalsatz*). As an introductory example, he presented Fig. 1 (shown on the previous page). It depicts the tripartite archenteron of a normal sea urchin pluteus (a), and two structures developed from the lower part of a gastrula after the upper half of both ectoderm and archenteron had been cut away; apparently, differentiation "occurred nonetheless essentially in typical proportions".

From this and similar examples, including isolated sea urchin blastomeres, regenerating *Tubularia* and sporulating *Acrasiales* (!), he derives the concept of harmonious equipotential systems. They are called equipotential because their constituent cells or groups of cells are claimed to be equivalent in their developmental potential – a rash claim hailing from 1893 (see Essay 10) to which Driesch was to cling stubbornly, despite mounting evidence to the contrary. The systems were called harmonious because their constituents when isolated will form a (smaller) pattern with identical proportions (harmonic proportionality; see Fig. 1). But whereas later researchers drew much benefit from his concept on a mechanistical basis, Driesch set out to prove that harmonious-equipotential systems cannot possibly work mechanistically and therefore constitute proof of (ultimately transcendental) vitalistic forces at work. "We have shown that in the realm of ontogenetic differentiation ... we are required to accept a new kind of elementary ... process, and we are so compelled because principally no machine produced by chemical and physical means can be contrived as the basis for the [patterning] events in question" (Driesch 1899, p. 99). In this veritable revolution, he went so far as to revoke his beautiful proposal that position acts via distance from sources of inductive stimuli (see Essay 11); he declared this proposal an illusion (Driesch 1899, p. 66) and dropped with it almost all theoretical considerations in the Analytical Theory!

Driesch's vitalism culminated in the Gifford Lectures. We have recourse to these only to show, using Fig. 14 on this page, that the problem of ontogenesis and especially pattern formation was really at the basis of his vitalistic reasoning, and that his "First Proof of Vitalism" (thus the heading of the corresponding chapter) was argued *per exclusionem*: "If we are going to explain what happens in our harmonious-equipotential systems by the aid of causality based upon the constellation of single physical or chemical factors and events, there *must* be some such thing as a machine. Now the assumption of the existence of a machine proves to be absolutely absurd in the light of the experimental facts. *Therefore there can be neither any sort of a machine nor any sort of causality based upon constellation underlying the differentiation of harmonious-equipotential systems*" (Driesch 1908, Vol. 1, p. 141; his emphases).

Logically, establishing proof by elimination is a tricky enterprise. Edwin Conklin (1929, p. 31) brought this out when quipping at Driesch's reasoning: "But this may mean no more than that the living machine is more complex than any that Driesch has in mind."

References

Conklin EG (1929) Problems of development. The Seventh Sedgwick Memorial Lecture, New York
Driesch H (1894) Analytische Theorie der organischen Entwicklung. Engelmann, Leipzig
Driesch H (1896) Die Maschinentheorie des Lebens. Biol Centralbl 16:353–368
Driesch H (1899) Die Lokalisation morphogenetischer Vorgänge – Ein Beweis vitalistischen Geschehens. Arch Entwicklungsmech Org [Roux' Arch] 8:35–111
Driesch H (1908) The science and philosophy of the organism, vols 1, 2. Adam and Charles Black, London
Driesch H (1951) Lebenserinnerungen. Aufzeichnungen eines Forschers und Denkers in entscheidender Zeit. Reinhardt, Munich Basle
Kant I (1790) Kritik der Urtheilskraft. Lagarde and Friederich, Berlin
Lenoir T (1982) The strategy of life. Teleology and mechanics in nineteenth-century German biology. D. Reidel, Dordrecht and Boston
Oppenheimer J (1971) Driesch, Hans Adolf Eduard. In Gillispie JJ (ed.) Dictionary of Scientific Biography, vol 4. Scribner, New York, pp 186–189
Spemann H (1927) Hans Driesch zum 60. Geburtstag. Roux's Arch Dev Biol 111:1–2

All translations by the present author, with explanations set in square brackets. Figure 1 is from Driesch 1899, Fig. 14 from Driesch 1908 (kindly provided by Lewis Wolpert).

ESSAY

Landmarks in Developmental Biology (13)

Entelechy and the ontogenetic machine – work and views of Hans Driesch from 1895 to 1910

Klaus Sander

The previous Essay traced some epistemological stages along Driesch's road to neo-vitalism, a transgression that is singular in view of his previous merits (and recognition) as an experimental embryologist. Here we shall review his experimental work between the teleo-mechanistic *Analytische Theorie der organischen Entwicklung* of 1894 (see Essay 11) and the end of his career as an active embryologist, which came shortly after his neo-vitalistic *Gifford Lectures* of 1908.

In these lectures, Driesch justified his vitalistic interpretation of ontogenesis, and of life in general, by what we may call the machine dilemma, his claim that in the face of experimental evidence it is impossible to ascribe ontogenesis to an "automatic machine" although from the mechanist's point of view there *must* be a machine in the egg (see Essay 12). While Driesch continued to accept mechanistic interpretations for subordinate steps in ontogenesis, he entrusted the task of their purposeful spatial and temporal coordination to a novel agent. Although immaterial and not resident in space, this agent was claimed to continuously act *into* space. Driesch characterized this view as "dynamical teleology", contrasting it with the "statical" teleology inherent in the developmental machine of his earlier writings (see Essay 12 and Churchill 1969).

In search of a respectable name for this new agent, Driesch (1901) alienated the Aristotelian term *Entelechy* – a term which in its new meaning was to become the hallmark of his neo-vitalism. When looking for a definition, the biologist may learn that "Entelechy has ruled the individual morphogenesis of the generation which is regarded as the starting point for inheritance, and will rule also the morphogenesis of the generation which is to follow" (Driesch 1908, Vol. II, p. 227/228). Or, phrased for the layman: "Entelechy, the locomotive engine driver, by way of repairs has the ability to influence at *any* time in *each* of its parts the engine subordinate to him, although often and for large stretches leaving it free to run by pure machine activities" (Driesch 1929, p. 6). And for the philosophically inclined: 'To build up the organism as a combined body of a typical style is the task of entelechy: entelechy means the faculty of achieving a "forma essentialis"; being and becoming are united here in a most remarkable manner: time enters into the Timeless, i.e. into the "idea" in the sense of Plato' (Driesch 1908, Vol. I, p. 149).

After these enlightening definitions we shall leave Entelechy alone, concluding with Wilhelm Roux that "since it is presumed to rule over material structure like a sovereign, this entelechy is no concept applicable to exact natural science" (Roux 1912, p. 241). And we cannot blame Driesch's old friend Morgan for lamenting (obviously with reference to Driesch) in his Nobel Prize lecture of 1934 that experimental embryology "unfortunately took a metaphysical turn".

Yet while moving towards his much-deplored vitalism, Driesch continued to produce research papers that caught the attention of his mechanist colleagues. In the period under review, he published – in addition to periodical reviews and dozens of writings on other topics (see Mocek 1974) – about 30 developmental papers. Among these we shall look here only at one of the earliest, and at the very last; we deliberately exclude Driesch's observations on dispermic sea urchin eggs, probably his most consequential work next to isolating blastomeres, because they relate to the topic of a later Essay. Finally, we shall turn to the question of what experiences or prejudices could have blinded Driesch towards mechanistic concepts capable of circumventing the machine dilemma – concepts that were demonstrably available by the turn of century (see the next Essay).

From among Driesch's publications of 1895 we pick out his joint studies with Thomas Hunt Morgan on comb-jelly (ctenophore) development, carried out at the Stazione Zoologica in Naples where Driesch played a crucial part in converting Morgan to experimental research (Allen 1978). Together they found that removing blastomeres from cleavage stages and removing cytoplasm from uncleaved egg cells could lead to the same types of defect, best defined by the absence of some of the eight "combs" and an incomplete internal organization causing the main axis to bend (see the Figures on this page). "From this fact we infer a localization of formative conditions in the cytoplasm of the ctenophore egg, and at the same time the absence of regulative powers in it." The opinion, so far founded only on indirect evidence, that the stimuli for embryonic differentiation come from the cytoplasmic structure of the egg cell "is directly proven by the results communicated herewith, and this at the same time deprives the Roux-Weismann theory of its last support, namely a certain logical justification that hitherto could not be denied. For while formerly one could only say: the cause for a typical differentiation [to occur] in a typical place must reside in the typical structure of the cytoplasm because it cannot rest upon [unequal] nuclear division, [...] we now can claim that it must reside in the cytoplasm because the removal of certain portions of cytoplasm is followed by the absence of typical organs in a certain place" (Driesch and Morgan 1895, p. 220/221).

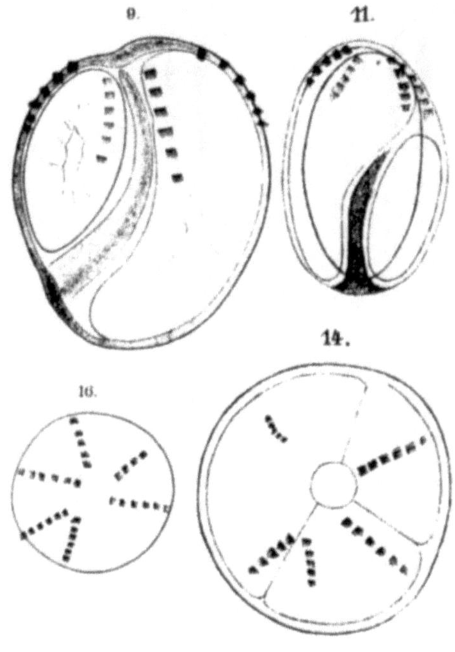

Driesch's last experimental work, published in 1910, was to show that fertilized egg cells when fused pairwise can give rise to uniform giant plutei. He had tried his hand at this experiment very early in his career, but he did not succeed until 1900, after Curt Herbst had shown him how to make the egg cells sticky by (inorganic) chemical tricks. While his shaking experiments (see Essay 10) had proven that one-half of an egg cell can produce a whole larva, he now demonstrated that a whole egg cell can be induced to form half a larva. In response to a critical comment by Boveri, Driesch in 1909 repeated the experiment, making sure that the egg batches he used were free of the oversized eggs that some females shed with low frequency – eggs which under the conditions of the earlier experiment might have given rise to the few uniform giant larvae he observed. The results, among them the giant gastrula and embryo shown together with normal controls on the next page, were reported in the *Festschrift* dedicated to Wilhelm Roux on his 60th birthday – surely a worthy setting for the last experimental paper of a disciple of *Entwickelungsmechanik* turned philosopher!

Having listed some of Driesch's achievements, we now turn to his limitations; fortunately, we can fall back on contemporaneous testimony, which should reduce the risk of unfair judgement based on hindsight. Our chief witness is Theodor Boveri, whose views on the ontogenetic machine in the sea urchin egg will be discussed in the next Essay. Suffice it to say here, in summary, "that the simple [spatial] differentiation of the cytoplasm serves to set in motion the machine whose essential and probably most complex mechanism rests in the nuclei" (Boveri 1902, p. 84/85). As we shall see, Boveri's concept owed much to Driesch's proposals in the *Analytische Theorie* (see Essay 11). However, it differed in one crucial point, for Driesch had dismissed the possibility that the nucleus harbours a "most complex mechanism".

Driesch's misjudgements may hail partly from the fact that he tended "to look at his objects from far above" (as suspected by Boveri in a letter to Spemann), but they may as well result from a deeply engrained preconception against any kind of invisible ordered structure in the egg, whether complex or simple, and whether in the nucleus or cytoplasm.

Such a preconception might be rooted in Driesch's epochal achievement, namely the abolition (at the age of 24!) of the neo-preformationist ideas of Weismann and Roux – ideas that had depended crucially on a complex but invisible structure preformed in the nucleus (see Essay 6). While kept at bay for some time with respect to the cytoplasm (see Essay 10), this antistructural inclination later on became underpinned by rational arguments concerning the nucleus, like the one that chromosomes appear homogeneous even under highest magnification, or that all chromatin particles react to a host of different stains "as a completely uniform substance, which at least does not favour considering them a chemical mixture, let alone a complex structure" (Driesch 1898, p. 66/67). That preconceptions played a role is evident in the case of cytoplasm, for Driesch in 1902 (p. 523) admitted to neglecting, in his "more essential theorizings", the possibility of preformed determinative structures in the cytoplasm – "because I thought of complete regulation of the disturbed relations" to occur after the elimination of vegetal cells.

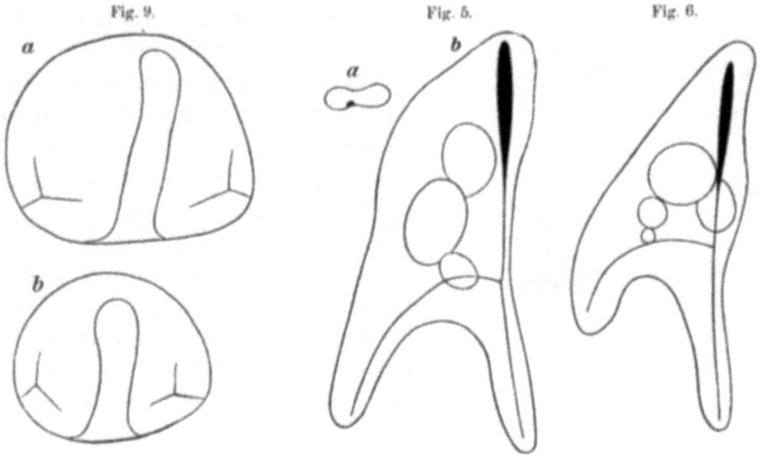

When Driesch finally, under the weight of arguments brought forward by Boveri, accepted in his embryological papers a specific role for the vegetal cytoplasm (Driesch 1902, p. 517), he nonetheless failed to embody this "causality based on constellation" in his philosophical considerations (where it might have endangered his "First Proof of Vitalism" outlined in Essay 12). Apparently, by then Driesch had passed the "point of no return" on his road towards vitalism, after which point the revocation of crucial arguments would have meant "loosing face". He also may have sensed intuitively that vitalism, in terms of social prestige (see Essay 12), promised to be more rewarding than its mechanistical alternatives.

However, it is only fair to state that at the time when Driesch *started* moving towards vitalism, the mechanists' claim that individual ontogenesis results from a "mere machine" was as much based on creed as was the vitalistic alternative; for mechanists even of the rank of Weismann und Roux blatantly failed to address, let alone to tackle, the grave problems posed by the assumed structural complexity within the egg cell nucleus. Driesch acknowledged these problems but sought non-mechanistical ways to eliminate them.

On this background, it is from *all* standpoints that Driesch deserves to be respected as the highly original experimentalist-plus-philosopher who, in the words of Julian Huxley (1927, p. 509), "perhaps more than any other single man has been responsible for the great advances of the last quarter of a century in the field of *Entwicklungsmechanik*; for he not only discovered a host of fundamental facts, but brought the whole subject into the centre of the biological stage by his insistence upon the vital importance of the theoretical questions associated with the facts, theoretical questions which his own powerful analysis helped to pose and to make precise."

References

Allen GE (1978) Thomas Hunt Morgan. The Man and his Science. Princeton University Press, Princeton NJ

Boveri T (1902) Über mehrpolige Mitosen als Mittel zur Analyse des Zellkerns. Verh Phys-Med Ges Würzburg NF 35:67–90

Churchill FB (1969) From Machine-Theory to Entelechy: Two Studies in Developmental Teleology. J Hist Biology 2:165–185

Driesch H (1899) Die Lokalisation morphogenetischer Vorgänge. Ein Beweis vitalistischen Geschehens. Arch Entwickelungsmech Org [Roux's Arch] 8:35–111

Driesch H (1901) Die organischen Regulationen. Leipzig (not seen)

Driesch H (1902) Neue Ergänzungen zur Entwicklungsphysiologie des Echinidenkeims. Arch Entwicklungsmech Org [Roux' Arch] 14:500:531

Driesch H (1908) The science and philosophy of the organism, vols. 1, 2. Adam and Charles Black, London

Driesch H (1929) Zur vitalistischen Begriffsbildung. Roux's Arch Dev Biol 116:1–6

Driesch H, Morgan TH (1895) Zur Analysis der ersten Entwicklungsstadien des Ctenophoreneies, I, II. Arch Entwickelungsmech Org [Roux's Arch] 2:204–215, 216–224

Huxley J (1927) The modification of development by means of temperature gradients. Roux's Arch Dev Biol 112:480–

Mocek R (1974) Wilhelm Roux – Hans Driesch. Fischer, Jena

Roux W (1912) Terminologie der Entwicklungsmechanik der Tiere und Pflanzen. Engelmann, Leipzig

All translations by the present writer, with additions set in square brackets. The illustrations on the preceding page are from Driesch and Morgan (1895), with Figs. 9 and 16 (reduced by 1/3) showing embryos deprived of a blastomere, while Figs. 11 and 14 resulted from the removal of egg cell cytoplasm; the other figures are from Driesch (1910) (reduced by 1/3), the inset (5a) locating the blastopore in the dumbbell-shaped blastula resulting from egg cell fusion that lead to the giant pluteus (Fig. 5b). Like the illustrations in previous Essays, these figures were skilfully reproduced from the originals by Gudrun Mahlke. The quotations from Driesch 1908 are taken from a copy kindly procured by Lewis Wolpert.

ESSAY

Landmarks in Developmental Biology (14)

Theodor Boveri on cytoplasmic organization in the sea urchin egg

Klaus Sander

Theodor Boveri, in the words of Edmund B. Wilson (1918), "throughout his life held with extraordinary tenacity to the investigation of a group of problems that were already clearly in his view during his first researches. What is the egg, and what does it contain? How are hereditary traits represented in the egg and determined during its development? What is the background from which they emerge into view, one after another, as development goes forward? By what mechanism (if mechanism it be) do the unseen potentialities hidden within the germ become realities that appear before our eyes in the embryo and adult?"

The impact of Boveri's quest and of his personality is revealed by two books that profoundly influenced the course of developmental biology in our century, *"The Cell in Development and Inheritance"* by Wilson (1896) and Hans Spemann's *"Experimentelle Beiträge zu einer Theorie der Entwicklung"* (1936; English edition 1938). Both books were dedicated by their authors to Theodor Boveri and his memory.

Although Boveri is best known for his work on nuclei and chromosomes, in this Essay we shall compare his concepts and those of Hans Driesch (see the previous Essays) on the cytoplasmic structure of the sea urchin egg and its contributions to embryonic patterning, as inferred from observation and experiment.

Theodor Boveri (1862–1915) was born in Bamberg, the son of a gifted but somewhat spendthrift physician (see Baltzer 1962; Oppenheimer 1970). The name Boveri is of Savoyan origin, but the family had lived for many generations in this part of Bavaria, or rather Franconia as the locals would have preferred to say (they came under Bavarian rule only some 60 years earlier, by the grace of Napoleon); the family name lives on in the industrial empire founded in Switzerland by one of Boveri's brothers and his English companion. Compared with Driesch's, the life of Theodor Boveri was utterly uneventful. He went to school near his birth place, studied anatomy (after a brief spell in history) in Munich, the nearest university of reputation. After his doctorate he stayed in Munich but moved to the zoological institute, where he achieved his Habilitation under the auspices of Richard Hertwig (the most influential younger brother of Oscar).

At the age of thirty, Boveri accepted the position of full professor and head of the Zoological Institute at Würzburg (again in Franconia) where – declining top offers from other universities and from the newly founded *Kaiser-Wilhelm-Gesellschaft* – he lived until his untimely death. In 1897 he married his American PhD student Marcella O'Grady; their only child, Margret, worked for some years as a secretary at the *Stazione Zoologica* in Naples and thereafter became internationally known as a political correspondent and writer. Boveri's travels abroad were restricted to the Alps, the French Riviera and Italy, save for a visit paid to London during the Darwin celebrations of 1908. Nearly a dozen times he spent the winter or spring working at the *Stazione Zoologica*, where he sometimes had to share the laboratory and the *pensione* with Curt Herbst and Hans Driesch, benefiting from the technical expertise of the former (see Essays 11 and 13) and getting his stomach upset from dinner discussions with the latter.

It would be hard indeed to imagine personalities differing more widely than those of Boveri and Driesch. Driesch, at least in his younger years, had hardly finished an experiment before he rushed the result to press. Boveri, even while young, pondered over his results, sometimes for years, before lecturing on them to the local science association (in Würzburg the *Physikalisch-Medicinische Gesellschaft*, who printed the lectures in its *Verhandlungsberichte*), after which he might wait another couple of years before turning out a voluminous masterpiece on the topic concerned; he did not write a single book, while Driesch published about ten before devoting himself fully to philosophy. Driesch's health and self-confidence were practically unfailing, whereas Boveri ailed for most of his life and more than once felt that he could never live up to the expectations that others placed in him.

In our discussion of cytoplasmic organization in the sea urchin egg, we shall start, as Boveri (1901 a) did, from the concept that Driesch had set forth in the *"Analytische Theorie der organischen Entwicklung"* in 1894. In his reasoning, Driesch had to struggle with a handicap pointed out by Wilson in his Woods hole lecture of 1893 (see Moore 1972), namely that "the echinoderm egg shows no visible differentiation in the cells until a relatively late period (16-celled stage)". Differentiation in the form of polarity could transiently be recognized by such landmarks as polar bodies, micromeres, and the archenteron, but Driesch had no means of knowing whether the polar axis thus marked is always the same; this is strikingly documented by his belief, held for many years, that micromeres and blastopore formed at *opposite* poles (see Essay 10). Consequently, he also lacked reliable criteria for identifying the animal or vegetal derivation of blastomeres isolated by shaking 8-cell stages, thereby missing the fact that animal blastomeres do not gastrulate (he may have taken for animal some (in modern designation) vegetal fragments that had failed to form the micromeres, his only criterion). Finally, and perhaps most important, Driesch could not know whether or not his experiments were altering the main axis of egg cell.

The uncertainties concerning axial organization came to an end in 1900, when Boveri discovered the pigment ring of *Strongylocentrotus* (now *Paracentrotus*) *lividus*, which has featured prominently since then in all textbooks. As shown in Figs. 3–8 on the front page, Boveri (1901 b) found that the ring did not form until after egg maturation, by aggregation of pigment granules that were initially evenly distributed, and that it was invariably oriented at right angles to the axis marked by the polar bodies. The location of these in turn was presaged by the germinal vesicle and yet another structure, the funnel-shaped pore penetrating the jelly coat. Using these additional polarity markers, and especially the pigment ring, Boveri could confidently state that axial polarity was maintained from oogenesis through fertilization and cleavage up to gastrulation and beyond, and that the micromeres and the archenteron invariably arose from the vegetal pole region, even in eggs strongly deformed by experiment.

These findings had far-reaching consequences for attempts to interpret the experimental results amassed meanwhile. Driesch's interpretation, as indicated in Essay 13, drifted from favouring preformed organization towards its denial. In the beginning, he had an inkling that the preformed axis was both stable and important for subsequent development. For instance, on discussing the development of 1/2-blastomeres he preferred to have the order of cell tiers maintained (see Essay 10). He also envisioned specific functions to be localized in the polar cytoplasm: "It is only because [a cleavage nucleus] comes to occupy a specific portion of the egg cytoplasm (whether in normal development or owing to experimental interference) that it becomes the nucleus of a micromere" (Driesch 1893, pp. 229/230). But by 1894, after noticing how little effect aberrant cleavage had on the outcome of development, he

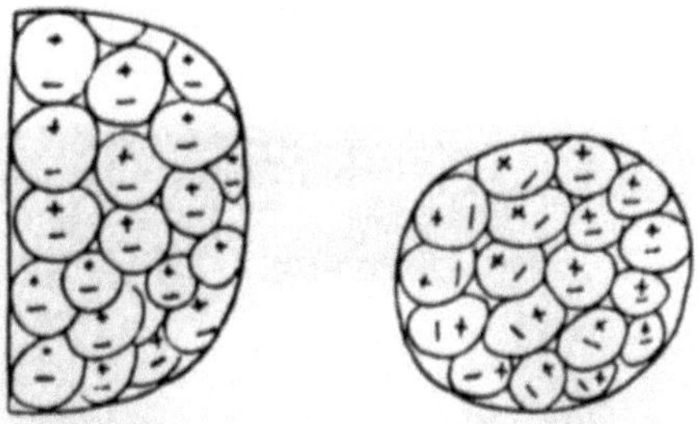

Fig. 1. Fig. 2.

had abandoned this concept in favour of polarity being handed down to each blastomere, with overall polarity resulting not from regional differences in the egg cytoplasm but rather (by analogy with magnetic polarity) from the near-identical orientation of polarity in all cells. Figures 1 and 2 above show a consequence of this for the development of 1/2-blastomeres: overall polarity, still maintained in the open half-blastula (Fig. 1), would be upset when this closes (Fig. 2, see also Essay 10). Overall polarity would then need to be re-established (in more or less random orientation) by a time-consuming and ill-understood process of cellular realignment; failure of this process would cause defective development.

Boveri rejected Driesch's concept on account of several shortcomings (Boveri 1901a). Not surprisingly, his own tenet was that the egg cytoplasm must be layered. In support of this, he lists the fact that the nascent pigment ring adopts a predictable position, as do the mitotic spindles during first and second cleavage; they align constantly just above the ring, in what he called the "karyokinetic plane". As for function (p. 160), "this layering into at least three zones should primarily aim at paving the way for the different qualities of the future primitive organs", namely micromeres (mesenchyme), archenteron and ectoderm (which Boveri erroneously derived only from the unpigmented animal half). It may be of interest to note that on discussing (and rejecting) this idea, Driesch (1902, p. 527) aspired to clarify it in terms of *polar coordinates*!

The assumption of cytoplasmic layering, however, by no means implied a mosaic status for the egg. Rather, it formed the basis of Boveri's ontogenetic machinery, in many ways the archetype of present-day concepts (see Davidson 1989). Boveri's arguments concerning the ooplasm (Boveri 1901a) deserve a closer look. To begin with, he explains (p. 165) that Driesch's concept could work only if a certain (polar) location were to act as a "centre of the first differentation from where all other parts have to be determined in their roles." This, however could be achieved equally well or even better by a layered cytoplasm with a "privileged" vegetal cap. 'If we assume that differentiation always starts at the "most vegetal" location', regulative development after various sorts of experimental interference can be understood because "each egg fragment will have such a relatively most vegetal location." Referring to the uniform plutei that Driesch (1896, p. 115) had obtained after severe scrambling of the blastomeres of an 8-cell stage, Boveri then suggests (p. 167) "that each region of the blastula is ready to form mesenchyme or to gastrulate, and that the localization to a single point is effected because these processes begin easier in this [most vegetal] region than in all others." After exempting the most animal region from this rule, he continues on p. 169: 'Should further experiments support this concept, then Driesch's famous sentence: "Each part, if need be, can do everything" should be modified to say: "Each part can do that, provided the one which could do it better is no longer present". The one that can do best, however, is probably the [relatively] most vegetative.'

What these assumptions amount to is postulating a graded vegetal property which at the site of its relative maximum serves as the first centre of induction. Boveri's student Hans Spemann, according to his testimony (Spemann 1936, pp. 91/92), was unaware of this idea when developing his own concept of the organization centre, but he willingly admitted that the idea of his great teacher might have been working unnoticed in him all the time.

References

Baltzer F (1962) Theodor Boveri, Leben und Werk. Wiss Verlagsges Stuttgart (English edition published 1967 by University of California Press, Berkeley)

Boveri T (1901a) Über die Polarität des Seeigeleies. Verh phys-med Ges Würzburg NF 34:145–176

Boveri T (1901b) Die Polarität von Oocyte, Ei und Larve des *Strongylocentrotus lividus*. Zool Jb Anat Ontog 14:630–653

Boveri T (1902) Über mehrpolige Mitosen als Mittel zur Analyse des Zellkerns. Verh phys-med Ges Würzburg NF 35:67–90

Davidson EH (1989) Lineage-specific gene expression and the regulative capacities of the sea urchin embryo: a proposed mechanism. Development 105:421–445

Driesch H (1893) Entwicklungsmechanische Studien VIII. Mitt Zool Stat Neapel 11:226–232

Driesch H (1894) Analytische Theorie der organischen Entwicklung. Engelmann, Leipzig

Driesch H (1896) Betrachtungen über die Organisation des Eies und ihre Genese. Arch Entwicklungsmech Org [Roux' Arch] 4:75–124

Driesch H (1902) Neue Ergänzungen zur Entwicklungsphysiologie des Echinidenkeims. Arch Entwicklungsmech Org [Roux'Arch] 14:500–531

Moore JA (1972) Readings in heredity and development. Oxford University Press, New York

Oppenheimer J (1970) Boveri, Theodor. In Gillispie CC (ed) Dictionary of scientific biography. Scribner, New York, pp 361–365

Wilson EB (1918) Theodor Boveri. In: Roentgen CW (ed) Erinnerungen an Theodor Boveri. Mohr, Tübingen

All translations by the present writer, with additions set in square brackets. The figures marked 1 and 2 are from Driesch (1894). The other figures are from Boveri (1901b); in the text he mentions that sperm entrance through the funnel (as implied in Fig. 8) is the exception rather than the rule; the dashed lines in Figs. 7 and 8 highlight the variable excentricity of the pronucleus.

ESSAY

Landmarks in Developmental Biology (15)

Aneuploidy disrupts embryogenesis: Theodor Boveri's analysis of sea urchin dispermy

Klaus Sander

The title of this Essay embodies an anachronism, because the term aneuploidy was not coined until several years after Theodor Boveri's death. But that term would be meaningless if all chromosomes were functionally equivalent, and consequently no such term existed at the time when Boveri demonstrated the so-called individuality of chromosomes for the first time, by analysing dispermy in sea urchins. It is his exploitation of this "experiment of Nature" and its developmental implications that we shall now discuss, returning thereby from the role of the egg cytoplasm in development (see the previous Essay) to the roles that the nucleus, and more specifically the chromosomes, play. The work reviewed in this Essay and the next was, according to E.B. Wilson (1918, p. 74/75), Boveri's "crowning achievement, whether in respect to excellence of method or importance of result."

We cannot fully appreciate the conceptual progress achieved by Boveri without briefly reviewing previous ideas on the organization and informational content (again an anachronistic but indispensable term) of chromatin. It was known by the turn of the century that the number of chromosomes remains constant over successive mitoses, but the constant size differences between individual chromosomes were hardly noted, let alone perceived as significant; therefore, few people felt the need to think of individual chromosomes as being different from each other in any essential respect. Even worse, the most influential authorities of the late nineteenth century were far off the mark with their speculations on chromosomal organization. For instance, August Weismann (1892), in his epochal *Keimplasmatheorie* (see Essay 6), envisioned the chromatin content of the nucleus as organized, following Walter Flemming (see Essay 2), in a single filament that would break up into separate chromosomes (of equal size) only late in prophase (see Fig. A and B below). More important, Weismann considered *each* of the resulting chromosomes to contain the *entire* genome, and this not just once but in several copies (see his sketch on this page), which had accumulated from previous generations and were held responsible for whatever ancestral traits the individual might exhibit. On the basis of this concept, a small fragment of any chromosome should have been sufficient to direct ontogenesis – which implies that an aberrant complement of chromosomes certainly would not entail deadly risks for the embryo!

Initially, Theodor Boveri also held the view that "each of the chromosomes brought together in the egg cell [at fertilization] contains all chromatin qualities of the species" (Boveri 1904, p. 43). But sometime before 1902 he abandoned this view – owing (once again) to a lead from Hans Driesch. Looking back, we wrote in 1904 (p. 43): "For some years past, ... certain reservations had crept up in me because of the pathological development of dispermic eggs, as demonstrated specifically for the sea urchin by Driesch (1892). Driesch raised a considerable number of dispermic sea urchin eggs individually and noted that they all ended up as strongly pathological blastulae (so-

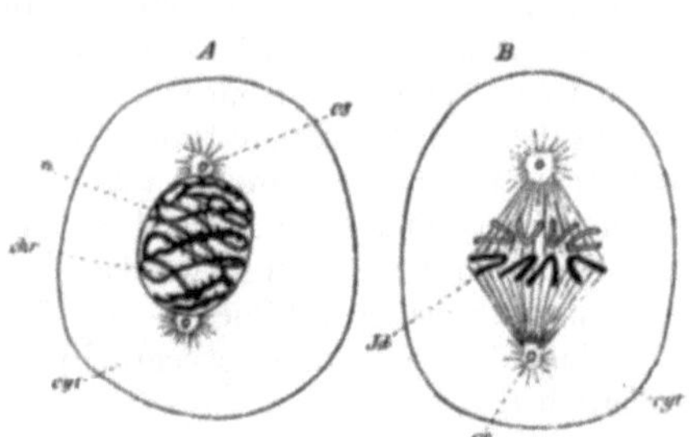

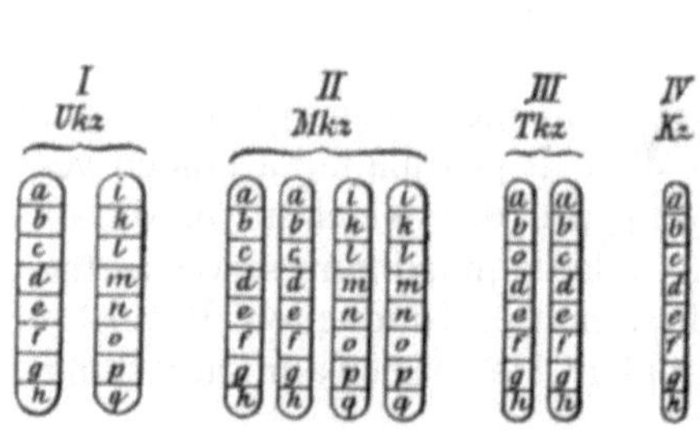

called stereoblastulae); not a single one was capable of gastrulation. On the condition that all chromosomes are equivalent, and taking into account all [previous] observations and experiments on echinid development, I was unable to conceive of a cause for this pathological development. So, when the discovery of Herbst (1900) had provided a method by which to separate sea urchin blastomeres safely and without damage, it was obvious that dispermic eggs should be used for checking on the problem of chromosomal equivalence" (Boveri 1904, pp. 44–45). So obvious was it that Boveri did not wish the news to spread: "By the way, I am asking you not to speak about the dispermy experiments for the time being. The matter is very simple, at least once one has hit on the idea of trying it" (from a letter to Hans Spemann; see Baltzer 1961 p. 119).

What Boveri failed to acknowledge in his recollections was one of Driesch's ingenious flashes of insight – in this case establishing a link between chromosome numbers and the fact that most dispermic sea urchin eggs produce a *Simultanvierer* at first cleavage. Driesch had written in 1892 (pp. 42–43), his emphases): "Now if an egg is fertilized by two spermatozoa, of which each carries along 1/2 of the [diploid] chromosome number, then we have 2/3 [to read 3/2] of them in the fertilized egg. *As this* [egg] *however divides simultaneously into four portions, each cleavage cell can receive only 3/4 of the normal chromatin mass*..., unless unequal distribution has made the anomaly even bigger. *I thus put the* [failing] *developmental capability of dispermic sea urchin eggs hypothetically on the account of insufficient nuclear material.*" Driesch (1892, p. 42) also stated his aim in publishing this speculation: "I wish just briefly to communicate a thought that may be helpful here, and the testing of which I leave to researchers experienced in cytology."

Boveri, second to none as a cytologist, evidently accepted this challenge. But, as was usual with him, he pondered over the problem for several years before announcing (in 1907) the ultimate solution, which we outline here by piecing together some sentences from a review he published when he was halfway through. At first he detailed the dilemma outlined by Driesch.: "In a dispermic egg the division of both sperm centrosomes as a rule gives rise to four [spindle] poles. The egg divides simultaneously into four cells. The question of interest to us is this: how are the chromosomes distributed among the four primary blastomeres? We shall assume for simplicity's sake a chromosome number of four in each pronucleus. Each of the twelve chromosomes will be arranged randomly between any two of the four poles. Fig. 48a represents one of the conceivable cases; Fig. 48b shows the subsequent stage after division of the egg into four cells. One can see that the chromatin content differs between the four blastomeres in number and kind; it is only in the lower left cell that all four kinds are present" (Boveri 1904, pp. 44–46).

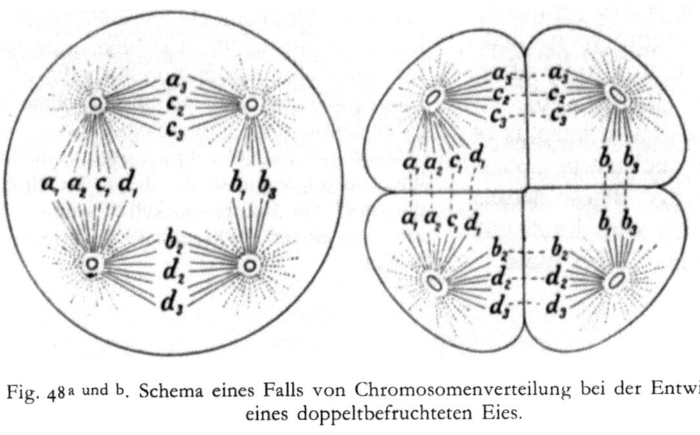

Fig. 48a und b. Schema eines Falls von Chromosomenverteilung bei der Entwicklung eines doppeltbefruchteten Eies.

From this followed a deduction which Boveri tested by separating the cleavage cells using Herbst's method: "The four cells arising by simultaneous division of a dispermic egg are essentially equivalent in all properties of their cytoplasm, including the centrosomes. If the pathological development of dispermic eggs results from a disturbance in the cytoplasm, then the offspring of all four cells must be aberrant in the same way; if it results from the anomalous chromatin complement, then one should expect that the [four] cells will differ [in their fates]. The experiments yielded the latter result in a most spectacular manner" (Boveri 1904, p. 47).

This all-important result is illustrated in Fig. 4a–d on the next page, but before turning to Boveri's explanations we should mention the control experiment: "Considering the normally fertilized egg, this is separated into four cells after two steps of division and then looks identical to the double-fertilized [egg] after its first division. If one detaches a single 1/4-blastomere of the normal germ from the three others, then it will yield, as shown by Driesch (1892), a normal but of course correspondingly smaller pluteus, which proof I was able to extend for our purposes [by showing] that all four blastomeres are capable of this in an identical manner" (Boveri 1904, p. 47). This then was – in 1901/1902! – the first demonstration ever of the quadruplets that Driesch (1892) had postulated, extrapolating from the development of single surviving 1/4-blastomeres.

This demonstration was needed for comparison with the blastomeres from dispermic eggs (cf. the Figure on the next page): "If however one separates the four blastomeres of a dispermic germ, something entirely different happens. Firstly they develop ... more or less pathologically; secondly, and this is the important point: the four blastomeres derived from a

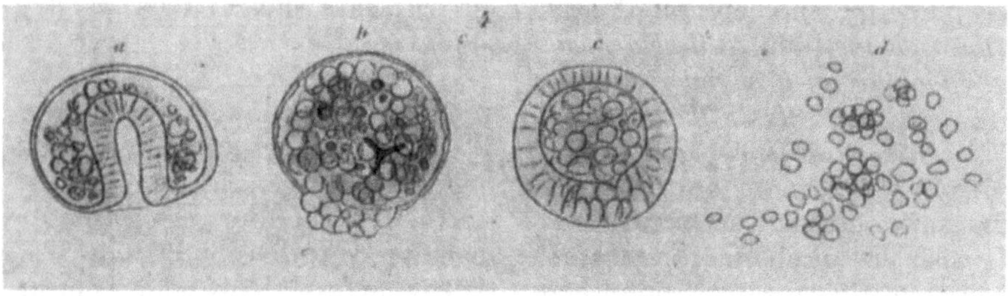

single egg and raised under identical conditions develop differently as a rule, and especially to different degrees. Development up to the blastula is almost always normal, but then differences appear: one quarter [of the *Simultanvierer*], for instance, decays into isolated cells, whereas the others swim about for several days as larvae; of these, one may perhaps remain in the blastula stage, a third gastrulates before ceasing development in this state, whereas the fourth may perhaps form spicules and show subdivision of its intestine, thus at least beginning its transformation into a pluteus" (Boveri 1904, pp. 47/48).

At that time Boveri was unable, for trivial reasons, to establish any chromosome numbers in his dispermic embryos, let alone to prove that it was the loss of specific chromosomes that condemned their blastomeres to fail in development. But he mustered indirect evidence along two lines. The first of these was based on his earlier finding that the surface area of a nucleus is roughly proportional to its chromatin content. He now noted that some of the isolated blastomeres had quite large nuclei but failed during subsequent development, while others with abnormally small nuclei did quite well. This finding excluded deficits in chromatin *quantity* from the possible causes of failure, thereby incriminating the absence of some chromosome(s) with a specific *quality*. "Thus what remains is that not a certain number, but a certain combination of chromosomes is required for normal development, and this cannot but mean that the individual chromosomes must possess different qualities" (Boveri 1902, p. 75).

The second piece of indirect evidence was statistical and underpinned by simulation, being based on developmental mosaicism, which will be related in one of the next Essays.

As is well known (witness the obituary by Wilson), Boveri's analysis of the "natural experiment" of dispermy was one of the cornerstones of the chromosome theory of inheritance: without chromosomal "individuality" – or, more clearly, the functional uniqueness of individual chromosomes – there would have been no point in paralleling the passage of the chromosomes through successive generations to the transmission of Gregor Mendel's hereditary factors. But this is exactly what both Theodor Boveri and Walter Sutton did almost independently of each other in 1902/1903, thereby laying the foundations of cytogenetics.

For developmental biology, Boveri's conclusions were potentially as important. For instance, he concluded – as suspected earlier by Driesch – that the embryo needed its chromosomal factors only after the blastula stage (Boveri 1907, pp. 253–254), and that a surplus of chromosomes can be just as detrimental as a shortage. These details, although interesting with hindsight, are but minor facets of his main message, namely, that a multitude of separable hereditary factors is required even for early ontogenesis. But this insight was largely lost on the generations of biologists succeeding Boveri, to be fully – and splendidly – revived only in our day.

References

Baltzer F (1961) Theodor Boveri. Leben und Werk eines großen Biologen. Wissenschaftliche Verlagsgesellschaft, Stuttgart. (English translation 1967, Univ of California Press)

Boveri T (1902) Über mehrpolige Mitosen als Mittel zur Analyse des Zellkerns. Verh phys-med Ges Würzburg [N F] 35:67–90

Boveri T (1904) Ergebnisse über die Konstitution der chromatischen Substanz des Zellkerns. Gustav Fischer, Jena

Boveri T (1907) Die Entwicklung dispermer Seeigel-Eier. Ein Beitrag zur Befruchtungslehre und zur Theorie des Kerns (Zellenstudien VI). Gustav Fischer, Jena

Driesch H (1892) Entwicklungsmechanische Studien III–VI. Z Wiss Zool 55:1–62

Herbst C (1900) Über das Auseinanderweichen von Furchungs- und Gewebszellen in kalkfreiem Medium. Arch Entwicklungsmech Org [Roux' Arch] 9:424–436

Weismann A (1891) Amphimixis oder die Vermischung der Individuen. Gustav Fischer, Jena

Weismann A (1892) Das Keimplasma – eine Theorie der Vererbung. Gustav Fischer, Jena

Wilson EB (1918) Theodor Boveri. *In*: Roentgen CW (ed) Erinnerungen an Theodor Boveri. JBC Mohr (Paul Siebeck), Tübingen, pp 67–89

All translations by the present author, with additions set in square brackets. The figures on the front page are from August Weismann, representing mitotic pro- and anaphase (1892) and chromosome behaviour in meiosis (1891). The subdivisions of the paired chromosomes do not indicate genes; each represents an entire genome (*Id* in Weismann's terminology); the sketches are part of a figure arguing for a linear arrangement of the Ids on the chromosome! The figures marked 48 and 4 are from Boveri (1904 and 1907, respectively).

ESSAY

Landmarks in Developmental Biology (16)

Reflections on method: Theodor Boveri's evaluation of "natural experiments" and their table-top simulation

Klaus Sander

As described in the previous Essay, Theodor Boveri, by exploiting the "natural experiment" of sea urchin dispermy, demonstrated that embryogenesis requires "a certain combination of chromosomes", possibly the full chromosome complement. The main topic of this Essay is his reflections on the methods he employed, and on their value for gaining previously inaccessible insights.

Since much of Boveri's sea urchin work started from the experimental and conceptual basis laid by Hans Driesch (see Essays 11 to 13), it is appropriate to begin with Boveri's reflections on his progress beyond Driesch's views. As may be remembered, Driesch the vitalist expounded the "machine dilemma" (Essay 13) which allegedly was besetting his mechanist opponents: in their view ontogenesis must be the outcome of a complex machine, whereas such a machine cannot possibly remain in working order if one removes or rearranges some of its parts (not to mention its duplication in twinning). This apory is rejected by Boveri: 'The contradiction constructed here by Driesch which, besides other considerations, had led him to postulate the "autonomy of life", to me does not seem to exist in reality. [...] I have shown earlier, and been able to prove more exactly since, that minor dislocations of cytoplasm at the vegetal pole lead to duplications, and I have meanwhile obtained from rearranged blastomeres [...] larvae with a duplicated and even triplicated archenteron and others with strong deformations and skeletal malformations. Accordingly, the echinid germ is nothing less than a harmonious-equipotential system. Finally, and that is decisive, one can take away randomly from the young echinid germ "any nuclei" (as shown by Driesch) but not "any part of the nucleus." To take away something from the nucleus, that has not even been attempted in Driesch's experiments; my own, however, which accomplish this, teach us that the nucleus, to which we now can ascribe any degree of complexity, behaves exactly as postulated for a "machine" by Driesch in his considerations' (Boveri 1902, p. 84).

The removal of parts from a cleavage nucleus (and thus from all its clonal descendants) was a welcome consequence of the multiple first cleavage division to which dispermic egg cells are prone. On considering the numerical consequences of dispermy for the chromosome complement of the ensuing blastomeres, Driesch was restricted to abstract reasoning (see the previous Essay). Boveri, however, could give numerical estimates, for he had shown, by investigations undertaken specifically to this end (Boveri 1905), that the surface of a nucleus is roughly proportional to its chromatin content. This allowed

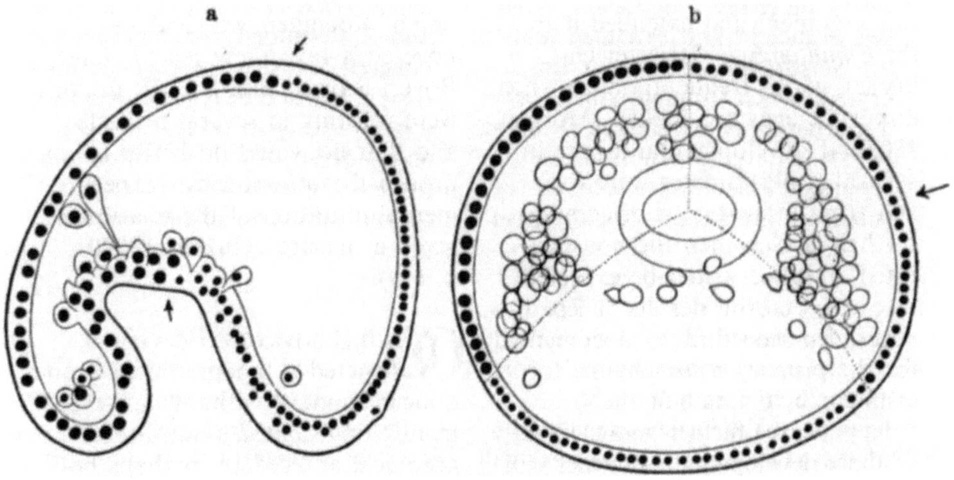

Fig. XXXVIII.

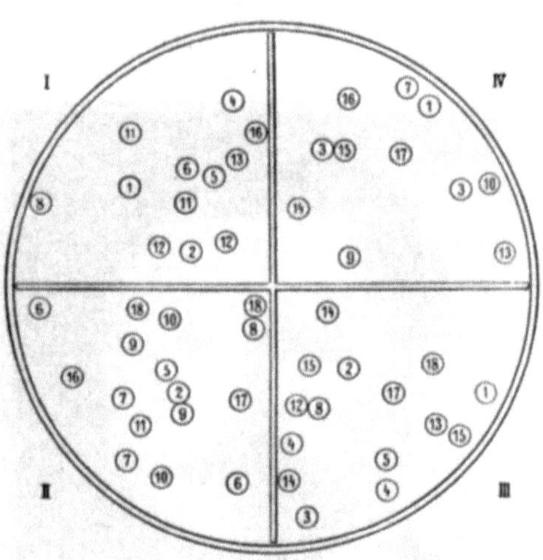

Fig. LXI.

Fig. LXII.

him to infer the approximate number of chromosomes in a nucleus from its diameter – a method that is extremely useful, because actual chromosome counts are much too tedious for analysing experiments that involved hundreds of embryos. The figure on the front page shows the clarity with which different nuclear sizes can be recognized even in the gastrula and later stages after hematoxylin staining and mounting in glycerol.

As may be remembered from the previous Essay, Boveri's first *experimentum crucis* was to separate the blastomeres after *Simultanvierer* – or tetraster – cleavage (see Fig. 48 in the previous Essay) and study their subsequent development. In his final paper (Boveri 1907), he tested a deduction from this argument and extended it to the *Simultandreier* (triaster embryos), which frequently arise when dispermic eggs are shaken (Morgan 1895). If developmental failure in separated blastomeres were to result from individual shortcomings of their nuclei, then the non-separated embryos would be expected to carry sectorial defects. These Boveri did indeed find, as documented for the primary mesenchyme, for example, by Figure b on the front page; the further implications of these developmental mosaics will be discussed in the next Essay.

Boveri's second *experimentum crucis* was a kind of experimental simulation. He had tried to calculate how frequently one should encounter dispermic embryos where every blastomere had received at least one chromosome of each kind, a condition which by now he considered indispensable for normal development. One could easily deduce in a general way that "healthy" embryos would be more frequent after triaster than after tetraster cleavage. Boveri's early calculations – assuming a haploid set of nine chromosomes (which number he had occasionally encountered) – indeed gave a yield of 4% for the *Simultandreier* and of 0.0026% for the *Simultanvierer* configuration. However, on discussing these efforts with the mathematician Wien (a mutual friend of his and of Roentgen, who had discovered X-rays a few years before), Boveri learned that his calculations were wanting in several respects and that he would do better to approach the question by "experimental imitation of the relations given in nature" (Boveri 1907, p. 149).

On this advice, Boveri constructed a topographical "random machine" (without, however mentioning Galton's non-topographical antecedent, perhaps because he did not know of it). It consisted of a cloth-covered circular board resembling a table top. On this flat "egg cell" he poured three "chromosome sets", each represented by a series of wooden balls numbered from 1 to 18 (the usual haploid chromosome number). This done, a wooden cross was placed on the board (see Fig. LXI on this page), thereby assigning a chance assemblage of "chromosomes" to each of the four "metaphase plates" usually appearing in tetraster cleavage (see Fig. 48a in the previous Essay). The balls were then transferred to a counting board (Fig. LXII) and the counts from pairs of adjacent columns (and also from columns I and IV) pooled to find the chromosome complement that each spindle pole would receive from the two metaphase plates joined to it. In an analogous way, chromosome distribution after triaster cleavage would be simulated by superimposing on the "egg cell" a triradius instead of the cross.

After performing 100 iterations for the *Simultanvierer* and 200 for the *Simultandreier*, Boveri asked how frequently every "blastomere" had received a full complement of "chromosomes" (i.e. at least one ball each of numbers 1–18). The frequencies of these "healthy" configurations were 11% for triaster cleavage and nil for tetraster cleavage. These figures received support from two additional results. (1) The percentages proved fairly constant when subsets of the entire sample, consisting of a smaller series of consecutive iterations, were checked. (2) The *Simultanvierer* simulation, besides not yielding any completely "healthy" cases, also yielded none with three healthy blastomeres and almost none with two (2%).

These frequencies were then compared with the percentages of apparently normal plutei actually developing from living dispermic eggs. After triaster cleavage, this class accounted for 8% among the living embryos as opposed to 11% in the simulation, a difference for which Boveri proposed several

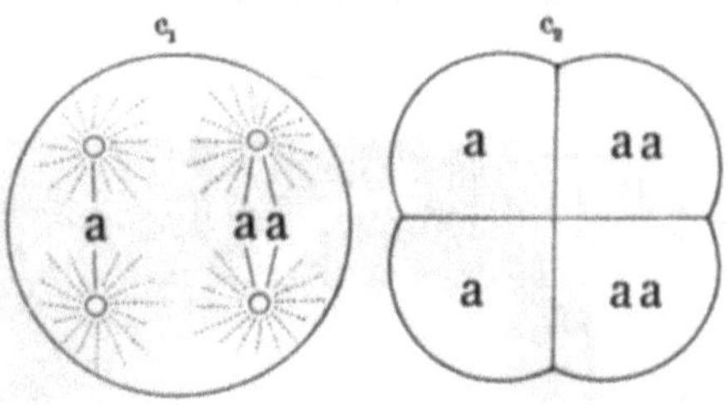

Fig. XXXIV.

plausible explanations. As for tetraster cleavage, the two normal plutei that Boveri (and his wife) found in mass cultures of more than 1500 *Simultanvierer* could also be explained, for instance by assuming the "double-spindle" configuration – one spindle dividing the zygote nucleus, the other the pronucleus developed from the second sperm. If followed (a very rare event) by simultaneous division into four cells, this configuration would yield two haploid and two diploid blastomeres (see Fig. XXXIV on this page) and thus should warrant normal – albeit mosaic – development, because haploidy is not known to disturb early sea urchin stages (see the next Essay).

Whatever confidence these explanations might command (see Moritz 1993), the essential result was that the percentage of normal "plutei" compared fairly well between the dispermic embryos and the "table-top simulation" of dispermy. Taken together with Boveri's other evidence, this agreement convinced most biologists that dispermy was detrimental because individual chromosomes would be excluded from one blastomere or another.

Some authors, however, thought otherwise (see Boveri 1907), among them Alexander Petrunkevitch, who rejected chromosome individuality owing to a mistaken view of meiotic segregation. Eduard Strasburger, on the other hand, declared that Mendel's laws proved chromosome individuality better than Boveri's experiments (which had actually been seminal for linking chromosome individuality to inheritance). Understandably, it was not without an undertone of irony that Boveri countered: "Both verdicts to me appear as conclusions from the less certain to the certain. But whatever one may think in this respect, I feel entitled to stress that my approach embodies real *experiments with chromosomes*, the results of which can be attacked only by rejecting the accuracy of the facts claimed or by disproving the conclusions drawn" (Boveri 1907, pp. 218/219, his emphases).

Yet it was another claim that prompted Boveri to comment on the general value of his method. Some critics had suggested that dissection of metaphase plates by cutting the normal egg cell during first cleavage would be superior to dispermy for obtaining random combinations of chromosomes. Here follows Boveri's response (1907, p. 216): "When I was confronted in writing and speech, and on the latter occasion rather vehemently, with the [supposedly] decisive and exclusive evidential value of the cutting experiment, I could not avoid the feeling that these utterances rested either on insufficient insight into the phenomena of karyokinesis, or on a widespread misunderstanding of the essentials of an experiment. In the eyes of some experimenters, if there was no cutting or no introduction of some substance alien to normal conditions, nor an apparatus fabricated by a mechanic, there was no experiment. Whereas the essence of the experiment rests solely in knowing for sure that certain circumstances, otherwise constantly present, are modified in a defined manner in a given case. Who had effected the modification, the observer or Nature herself, is irrelevant. Indeed, the investigator of Life will especially endeavour to find deviations from the normal where he himself, with his crude techniques, has not interfered at all and yet is able to fully comprehend the kind of modification [that has occurred]. One such experiment of Nature is dispermy. What an experimenter might wish to accomplish by cutting the egg is here achieved in unsurpassable ways. With complete certainty we can state that multipolar spindles will carve the daughter nuclei from the available chromatin complement as randomly with respect to number and combination of chromosomes as if we were to cut the nucleus with a knife. The delicacy, however, with which the karyokinetic machinery accomplishes this feat is infinitely superior to the knife's cut."

Which brings to mind the recent stupendous success of genetical methods in embryological research, a discipline for long dominated by the knife.

References

Boveri T (1902) Über mehrpolige Mitosen als Mittel zur Analyse des Zellkerns. Verh Phys-Med Ges Würzburg (NF) 35:67–90

Boveri T (1905) Über die Abhängigkeit der Kerngröße und Zellenzahl der Seeigel-Larven von der Chromosomenzahl der Ausgangszellen (Zellenstudien V). Jena Z Naturwiss 39:1–80

Boveri T (1907) Die Entwicklung dispermer Seeigel-Eier. Ein Beitrag zur Befruchtungslehre und zur Theorie des Kerns (Zellenstudien VI). Fischer, Jena

Morgan TH (1895) A study of variation in cleavage. Arch Entwicklungsmech Org [Roux' Arch] 2:72–80

Moritz KB (1993) Theodor Boveri (1862–1915), Pionier der modernen Zell- und Entwicklungsbiologie. Gustav Fischer Verlag Stuttgart/Jena

All translations by the present author, with additions set in square brackets. The figure on the front page, an early gastrula derived from triaster cleavage and showing (in optical section) three classes of nuclear diameter, is from Boveri 1907, as are the other figures. Note that Fig. LXI, in order to match the corresponding representation in Fig. 48a of the previous Essay, should be tilted by 45°; the quadrants do not represent blastomeres but metaphase plates and their chromosomal content before duplication. the figure marked XXXIV shows the double-spindle type *Simultanvierer* and actually was intended to illustrate the distribution of just one individual chromosome, but the single letter can as well be taken to represent a whole haploid chromosome complement as done in the text.

ESSAY

Landmarks in Developmental Biology (17)

Lessons from a queen's lost offspring – Theodor Boveri on honeybee gynanders and other genetic mosaics

K. Sander

When it came to making his audience see a point, Theodor Boveri proved himself a master in the formulation of enlightening parallels. Take this one for instance, forging a link between the present Essay and the one before.

"Like the prism which resolves the light emanating from a body and thereby signals the latter's constituent substances, embryonic development resolves the qualities contained in the egg by generating from them the final state. The unfolded organism is thus like a spectrum by which the countless minuscule peculiarities of the germ cell are made accessible for our means of perception. Therein rests the limitless superiority of all experiments on nuclei relating to egg or embryonic cells – and these include hybridization experiments – over experiments dealing with adult cells or unicellular beings. For far and wide, no other cell provides us with such a fine measure of its inherent qualities as do the egg cell and its nucleus when they shape the complete organism. *Embryonic analysis of the cell nucleus* – thus one might characterize the method from which we can anticipate the best fruits in our newly planted field of research" (Boveri 1904, p. 124).

The amazing details of "nuclear constitution" that Boveri had inferred by means of this method were the topic of earlier Essays. By inverting this approach (and his phrase) we shall now practice *nuclear analysis of embryogenesis*, asking how Boveri analysed specific developmental functions by exploiting constitutive differences between the nuclei (and consequently cells) in mosaic individuals of both sea urchins and the honeybee.

As for sea urchin mosaics, Fig. XXXVIII in the previous Essay has shown how nuclear size – different owing to unequal chromosome distribution in triaster cleavage – can characterize and mark the clones derived from early cleavage cells. Compared with the sublime art of modern "clonal analysis" (see e.g. Gehring 1978), the developmental information that Boveri drew from his sea urchin mosaics remained scanty – if only because a pluteus is poorer in morphological landmarks than *Drosophila,* and their descent during embryogenesis may seem easier to grasp by direct observation. Nonetheless, Boveri in 1907 (see Essay 16) concluded for instance that primary mesenchyme cells from different (clonal) sectors tend to mix while some of the epithelial sectors may fail to attract them properly (pp 95, 127; witness Fig. XXXVIII), and

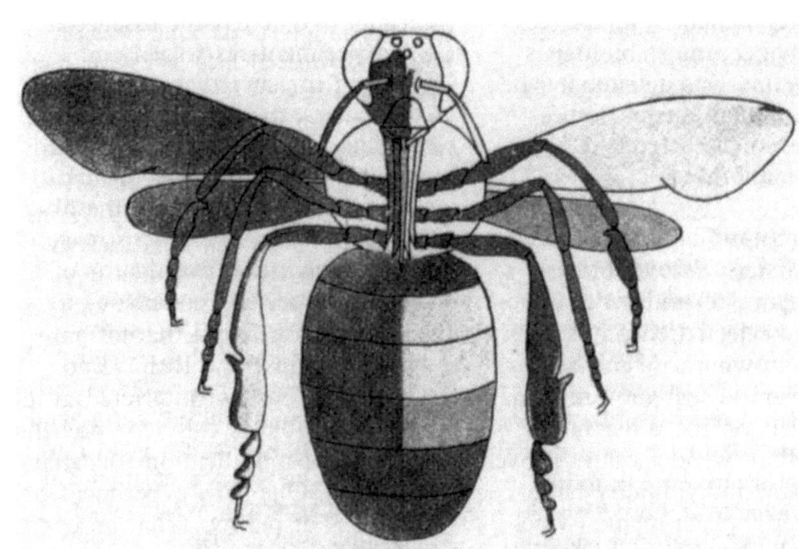

Fig. A

that distinct clonal sectors as a rule respect the larva's median plane (p 83).

Boveri's interest in honeybee gynanders, however, was kindled by euploid rather than aneuploid sea urchin embryos. His crucial finding – and one that came to dominate his reasoning (see the next Essay) – was "partial fertilization", the union of the paternal pronucleus with one of several haploid cleavage nuclei, giving rise to a haploid/diploid mosaic larva (Boveri 1888). He predicted in the same year that in the honeybee a similar course of events would lead to gynandromorphic individuals, a prediction relying on Joseph Dzierzon's claim that drones arise from unfertilized eggs.

Gynanders, individuals pieced together mosaic-fashion from female and male patches of integument (see Fig. A on the front page), had been known in insects and some other arthropods since the eighteenth century, when they generated much excitement among the savants. However, finds were few and far between, and it was not until 1860 that a rich source of material could be tapped. This source was a honeybee hive in Konstanz which had started pouring out gynanders in large numbers. I say "pouring" advisedly, because the worker bees flung these pitiful creatures out of the hive as soon as they emerged from their cells. Since they were unable to procure food, most gynanders perished on the ground below the hive – a lost and wasted progeny.

If lost to the hive they were not lost to science, because the hive's owner, one Mr. Eugster, maintained the colony for several years despite its poor returns, and brought it to the attention of zoologists in Zurich and Munich. Carl Theodor von Siebold came over from Munich, and it was he who in 1864 provided the first good description of honeybee gynandromorphism, which he based on a large number of Eugster's specimens. The one shortcoming of this work had to do with a clonal marker kindly introduced by Nature herself when she instigated the mating of the hive's yellow Italian queen with a dark German drone (see the next Essay).

After his investigations, von Siebold stored some of Eugster's *Zwitterbienen* in the collection maintained by his institute in Munich. Boveri, working there some 20 years later, was itching to lay his hands on whatever remained of Eugster's bees – but the curator told him that, alas, they were lost. After another 20 years, Boveri initiated a second search for these specimens. Favoured by a change in curatorship and possibly also helped along by Boveri's ever mounting reputation, this search brought back to light the doubly lost offspring of Mr. Eugster's queen.

Turning now to gynandromorphism as a developmental problem perceived by Bovery, we notice the striking absence of any direct reference to the cell-autonomous expression of sex characters in gynanders, stated so clearly by Morgan's group about the same time. However, on reading the paper published by Boveri's student Elsa Mehling (1915), one cannot fail to notice that they were both quite aware of the phenomenon but apparently took it for granted. To quote Mehling (1915, p. 205): "No doubt one has to assume that the mosaic expressed later on is defined essentially at a time when there is no indication as yet of the future organs. During subsequent differentiation the positioning of the organs will be determined only by the normal topographical relations, regardless of whether the cells that aggregate in a certain organ carry the male or the female sexual tendency." For instance the leg can carry proximodistal stripes differing in sexual character (see Figs. A and 56b), because "as soon as the male or female cells come to lie at [one] side of the bud, they will emerge as a narrow stripe in the adult leg" (p. 205).

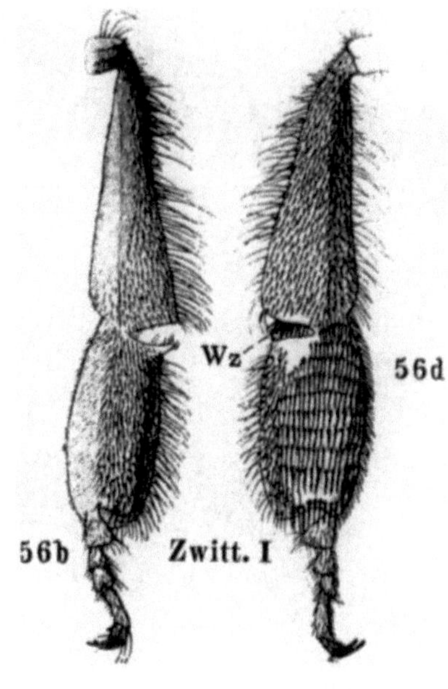

The opposite of cell autonomy, namely a certain coordination of male and female cells, seems to occur with respect to size. Mehling (1915, p. 196) notes this for instance in mosaic compound eyes. Drone ommatidia are both longer and wider than their (female) counterparts in workers. In mosaic eyes, the ommatidia of either sex, independently marked by differences in pigment distribution (see the histological section, Fig. 91), maintain their width over most of their repective domains, but both approach an intermediate length and thus provide continuity to the eye's surface contour.

Boveri in his own paper (1915), the last to appear during his lifetime, suggests that the random orientation of the generalized borderline between male and female parts is "a consequence of different orientations of the first cleavage spindle. As in the frog [⋯], it may also happen in the bee's egg that the first two [sexually differing] nuclei, and consequently also their descendants, may divide between them the right and left of the egg in one case, and in another its front and rear, and perhaps sometimes its top and bottom" (p. 299). The patchiness of the mosaic pattern visible in the adult would then (to some degree) result from shifts during subsequent growth and especially during metamorphosis,

91

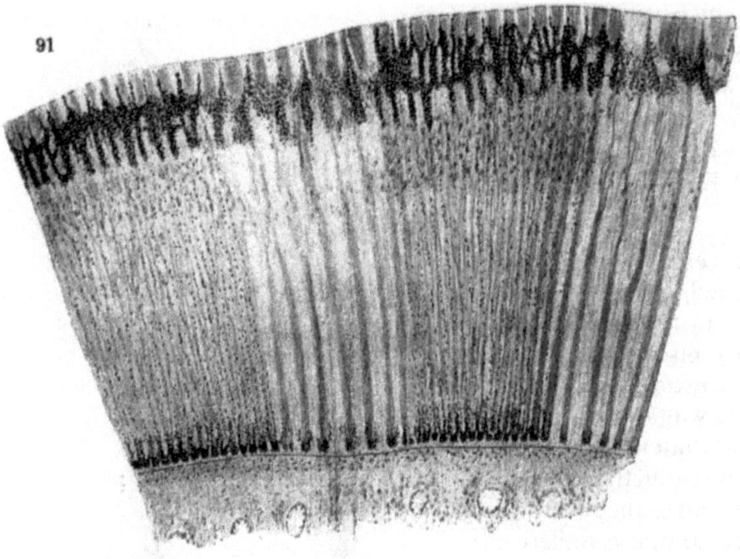

events which could influence a boundary's course either randomly or reproducibly. As an example for constantly occurring deflections of some non-random boundaries in the head, Boveri (p. 301) mentions protruding kinks like the one seen between the antennae in Fig. A.

Turning back to embryogenesis, Boveri draws attention to some rather common differences in arrangement and size between ventral and dorsal domains of adult gynanders. To readers remembering the striking, strictly median boundary on the back of silkworm mosaics (or some *Drosophila* gynanders), it may come as a surprise that in honeybee gynanders it is mainly the ventral side where these straight median boundaries occur, correlated with larger domain sizes than are found in more dorsal regions. Boveri (p. 303/304) concludes "that this contrast between the ventral and the other sides has its basis in the peculiar development of the insects, in which only part of the blastoderm proceeds to form the embryo while another part is eliminated as the embryonic covers [···]. These events probably comprise several steps that lead to strong mutual shiftings of cells. As far as I am aware, it remains uncertain whether [··· in the honeybee] the blastoderm cells aggregate towards the ventral side so that those remaining dorsally have to flatten and expand. If so, this would provide a first occasion for the separating of originally adjacent cell groups, and the rapprochement of [some others] initially separated. Similar, but more pronounced, effects should result from the upward growth of the germ band's free lateral margins and their dorsal fusion" – expectations which surely make sense in view of the extensive movements of cells and epithelia recognized in the honeybee embryo since then (see Fleig and Sander 1988).

Boveri's inferences from the adult mosaic back to its ontogenetic development, only some of which are mentioned here, were certainly phrased more cautiously and sometimes more vaguely than the detailed conclusions drawn in our time – not always correctly – from the analysis of genetic mosaics. Yet his closing words prove that he recognized the promise of this approach, though it was not realized until our day: "The yardstick by which we have to measure the theory of gynandromorphism cannot be our scanty knowledge of those embryonic growth processes; rather the reverse: like mosaicism in plant chimeras, by which we can fathom certain aspects of growth, the gynandromorphs – so similar in principle – will be able to bring clarity into some morphogenetic processes which cannot be unravelled by [observing] the normal events alone" (Boveri 1915, p. 304).

To which lesson he added, perhaps disheartened by his failing health: "But these considerations reach beyond the borders which I have set myself for this contribution."

References

Boveri T (1888) Über partielle Befruchtung. Sitzungsber Ges Morph Physiol München 4:64–72

Boveri T (1915) Über die Entstehung der Eugster'schen Zwitterbienen. Arch Entwicklungsmech Org [Roux' Arch] 41: 264–311

Fleig R, Sander K (1988) Honeybee morphogenesis: embryonic cell movements that shape the larval body. Development 104:525–534

Gehring WJ (ed) (1978) Genetic mosaics and cell differentiation. Springer, Berlin Heidelberg New York

Mehling E (1915) Über die gynandromorphen Bienen des Eugsterschen Stockes. Verh Phys Med Ges Würzburg 43:173–236

All translations by the present author; omissions and additions marked by square brackets. The figures are from Mehling (1915). Figure A is a scheme showing the overall distribution of male (*blank*) and female (*dark*) cuticular domains of a gynander closely studied by her; *lightly shaded parts* are of uncertain status. The largely worker-type (i.e. female) legs in Fig. 56 are both mosaic, (*b*) showing a male stripe at the plantar side (*left*) and (*d*) the suppression – by a male patch? – of the protrusion below the pollen press notch (marked *Wz*, cf. Fig. 56 b). Figure 91 is a section through a mosaic compound eye, showing the wide, less pigmented, male next to the slim female ommatidia.

ESSAY

Landmarks in Developmental Biology (18)

True alternatives: Boveri and Morgan arguing about the origins of honeybee gynanders

Klaus Sander

The preceding Essay was devoted to conclusions on embryogenesis that Theodor Boveri drew from genetic mosaics, mainly the famous gynanders of Mr. Eugster's honeybee hive. Here we shall return to the question of how gynandromorphism might come to be in the first place.

This question deserves some pondering not only because of the interest that gynanders and concomitant ideas on fertilization have always commanded, but also because two giants of developmental biology argued over it in a controversy that is remarkable for two reasons: its outcome and its noble style that differed so much from the polemics which Roux and Driesch indulged in at times. Moreover, the outcome of the conflict can teach a lesson too easily overlooked in the hot pursuit of competitive research, namely, that more explanations can be conceived than the opponents realize – and, perhaps even less palatable to them, that each of these alternatives can be true, especially in a context as complicated as ontogenesis.

Before turning to these conflicting arguments, let us have a look at von Siebold's opinion of 1864. Early on, he cautioned his readers: "Do not expect me to provide you with insights on *how* these hermaphroditic malformations come to originate. To render account in this question, much more experience has to be accumulated; indeed it may perhaps remain entirely open whether human acumen may ever succeed in grasping ... the circumstances of procreation which count among the most difficult problems in natural history" (p. 74). Towards the end, however, von Siebold could not resist the temptation to offer some speculations, based on Dzierzons theory that drones arise from unfertilized eggs, a theory published a few years earlier. Von Siebold concludes that "whereas in other animals the fertilizing male semen serves to initiate development as such, the influence of the drone's semen is directed towards the development of female individuals in eggs which per se ... are capable of development but generate merely male individuals" (p. 78/79). He then suggests that female development will most probably require a certain minimum number of sperm filaments. "If for whatever reason the egg's contents are not provided with the required number of sperm filaments, then the honeybee egg ... would not be able to achieve the production of a female bee, yet ... it would be disturbed in the parthenogenetic development of a pure drone in such a way that female traits of organization get mixed in, and by this incomplete fertilization the above-mentioned different degrees of sexually mixed forms come to arise. – On the impediments which cause the queen bee to incompletely fertilize eggs intended for yielding workers I can at present give no information" (von Siebold 1864, p. 79).

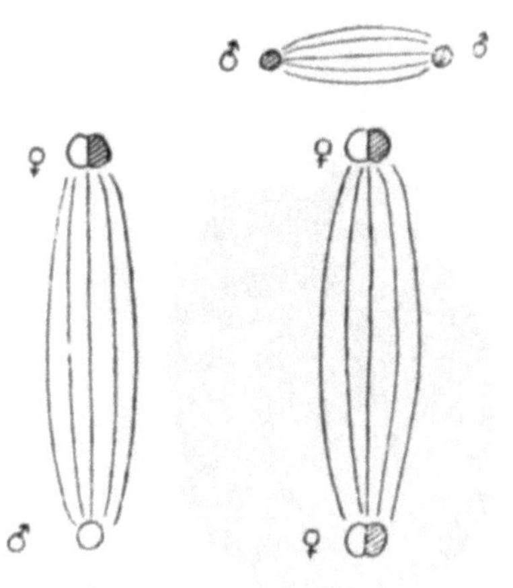

Half a century later, the divergent views of Boveri and Morgan of course rested on a different concept of fertilization, based foremost on Oscar Hertwig's claim that in sea urchin fertilization it is a single sperm, or rather its nucleus, that counts. For a clear-cut exposition of views and consequences we turn to Morgan (1905, p. 632ff; see the figures on this page): The view recently proposed by Boveri has met with the most favour .... Boveri has suggested that the gynandrous condition may be due to the spermatozoon failing to unite with the egg nucleus, but subsequently pairing with one of the products of the first division, as shown in diagram A. If a union of this sort should occur, all of the cells that are derived from the paired nucleus might be expected to give rise to female characters (as in the case of the *fertilized egg*); while all the cells that come from the unpaired half of the nucleus might be expected to produce male parts .... On purely theoretical grounds I offer an alternate hypothesis which will explain the facts equally well .... I suggest this alternation, especially as the two views can be put to the test of actual observation by any one in position to obtain the necessary material; and also because the possibility of this interpretation appears to have been entirely overlooked as an explanation of gynandromorphism. I suggest in brief that the results may be due to two (or more) spermatozoa entering the same egg, one only fusing with the egg nucleus, and the other not uniting, but developing without combining with any parts of the egg nucleus, as shown in diagram B."

As for consequences, Morgan continues: "It will be observed that on Boveri's hypothesis the male characters will be derived from the egg nucleus, while in my view they will come from the unpaired sperm nucleus. If, therefore, a queen of one race and a drone of another should produce one of these gynandromorphs we ought to be able to decide which of these views is correct; for on Boveri's hypothesis, the male characters of the gynandromorph would be those of the race to which the mother belongs, while on my view they would be those characteristic of the race of the father .... Von Siebold had in fact a case of this sort in the famous Eugster hive from which he obtained his material." However, as indicated in the previous Essay, Morgan was not happy with the use that von Siebold had made of this material: "Unfortunately his statement in regard to the racial characters of the gynandromorphs is obscure" (Morgan 1905, p. 633).

On this point Boveri (1915, p. 286) had a different opinion. Reading "between the lines" of von Siebold's rather unfocused text, and evaluating its messages in the light of contemporary knowledge, Boveri found von Siebold's statements compatible with his own observations to which we now shall turn.

On gazing at the rediscovered Eugster specimens (see the previous Essay), Boveri immediately noted patches of yellowish cuticle in their male domains. But whether this meant their descent from the maternal strain he could not decide because the existing descriptions of Italian bees were lacking the necessary details. Thus he sat down and carefully documented the cuticular markings of the pure-bred *Apis mellifica ligustica* (the queen's race) and the dark German strain. Returning to the gynander material, he realized yet another problem: might not the yellowish male domains lag behind in development – drones take three more days than workers until eclosion – so that they had not yet completed the darkening process? Luckily, by combining several observations Boveri could dismiss this doubt and convince himself of the maternal character of at least some male domains. The figures marked 1, 2 and 19a illustrate clear-cut proof: in several gynandromorphic head capsules, the half shown to be male by its large drone eye and the positions of its frontal ocelli had a brightly yellow clypeus – a typical *ligustica* trait. Convincing proof is provided (in Fig. 19a; see next page) by the lateral row of dark lunules standing out clearly in the male half (at the left) of a mosaic abdomen seen from below.

In summing up his findings, Boveri (1915, p. 291) declared: "The conclusion which ... we have to draw is short and simple. If among the two hypotheses of Morgan and myself, which are so similar to each other, any should be true at all, then the findings communicated here clearly speak in favour of mine." He continued (p. 292) by discussing a set of additional assumptions which would be required for interpreting his findings in Morgan's way – only to conlude that at least one of these assumptions (involving cytoplasmic inheritance) "is so absurd and so much conflicts with everything we know about the role of the nucleus in inheritance ... that no one would defend it, certainly least Morgan himself who by his experiments on inheritance has provided one of the finest arguments compelling us to localize the units of heredity in the chromosomes."

Morgan promptly admitted defeat – for the time being: "Briefly Boveri finds that the male parts of the gynandromorphs are maternal, while the female parts are paternal, which is the dominant character. This conclusion gives a decisive answer in favour of his hypothesis and sets my own aside for this case at least" (Morgan 1916, p. 40).

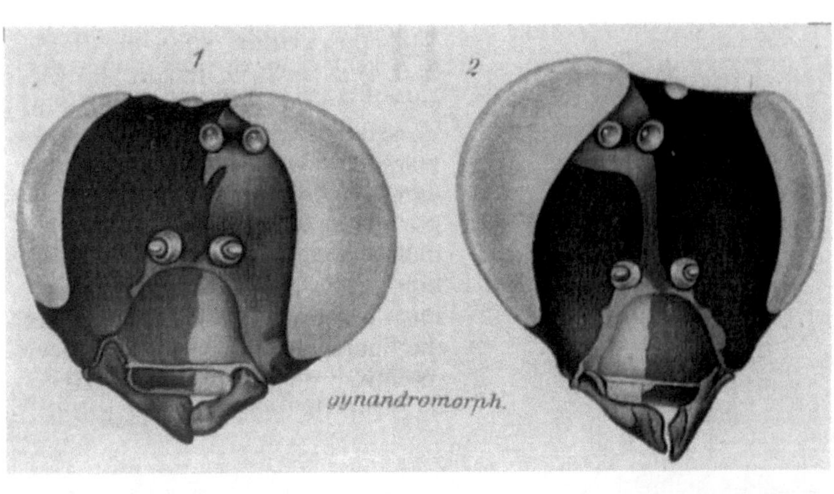

That Morgan, in restricting his declaration to "this case", was guided by prudence as well as facts is obvious from the subsequent pages of his paper. There (1916, p. 41) he cites a contemporary report on gynanders from the Caucasian region claiming paternal traits in the male domains and then takes recourse (p. 44) to certain silk moth gynanders. But, like Boveri (1915, p. 305), he considered the former claim to be based on insufficient evidence, and then observed that support by the silk worm data must be qualified: "If the same conditions prevail here as in the bee, viz., one nucleus producing a male and two producing a female, the case is in harmony with my hypothesis and not with that of Boveri. But the evidence for my view is not as strong as .... Boveri's is now for the bee; yet", Morgan continues, "it may be true, nevertheless, that in both of these ways gynandromorphs may arise". This prudent statement was followed, after a brief excursion into *Drosophila* gynandromorphism, by a less fortunate attempt at generalization: "In fact we should expect that gynandromorphs would arise in insects whenever certain nuclei come to contain two sex chromosomes and others only one". This notion later proved to be mistaken, as did his expectation (implied in the "whenever" and stated expressly in Morgan 1907, p. 718) that it is the quantitative differences between haploidy and diploidy which code for the two sexes – and not, as is now known, qualitative allelic differences at a sex locus (see Woyke 1963).

Both Morgan's fame as a geneticist and subsequent experimental proof involving marker mutants (see Milne 1986) caused widespread acceptance of Morgan's view that gynanders arise from accessory sperm nuclei. Does this mean that Boveri was wrong? For sure he was not completely wrong, to say the least.

Boveri's analysis of the cuticle markers in Eugster's *Zwitterbienen* was never challanged seriously, and subsequent research has shown that male domains carrying

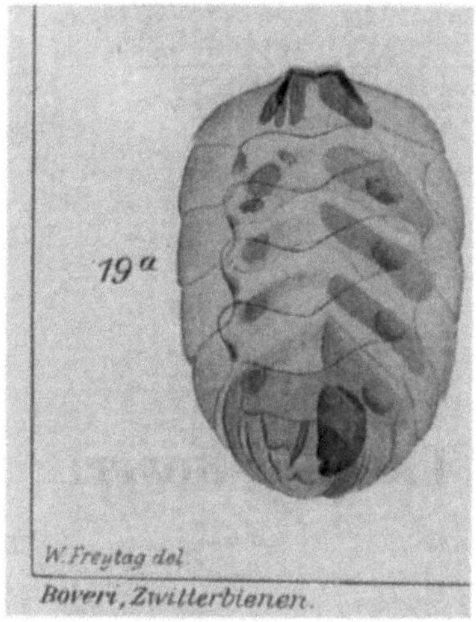

maternal markers can indeed occur both in the honeybee and in *Habrobracon* (the early star of hymenopteran genetics). However, in most instances it is proven or at least likely that these domains do not derive from a parthenogenetic cleavage nucleus, as envisioned by Boveri (see Fig. A on the front page), but rather from a "polar body" nucleus that failed to be inactivated (probably because of aberrant positioning). Interestingly, Morgan had considered this possibility for the silk moth (1907, p. 717), but not for the bee, and Boveri was probably prevented from seeing it because of his preoccupation with partial fertilization in the sea urchin (see Essay 17) and because of the role he had ascribed to the centrosome in fertilization (see Morgan's critique, 1916, p. 43).

However, again this does not necessarily mean that Boveri was wrong. Considering the incredible array of cytological anomalies during honeybee meiosis and fertilization known to be compatible with subsequent development (see Milne 1986), one must admit that he may have been right, especially since his specimens differed in several respects from all other honeybee gynanders with matrokline male domains: they were much more frequent, they were the only type observed in the hive (as long as the *ligustica* queen lived), and – last but not least – they came from a cross between subspecies while in all experiments involving marker mutants these were derived from a single subspecies or parental strain. Since the various anomalies shown to cause or facilitate gynandromorphism in the honeybee must all depend to some degree on maternal and/or paternal gene expression during gametogenesis, it is quite possible that a subspecies cross may have consequences not known from crosses within a subspecies.

In any case, Boveri's pioneering explanation of honeybee gynandromorphism enjoys less fame than Morgan's, although this, as we have seen, by no means represents the only alternative. From this, as Boveri (1915, p. 295) remarked in a different context, "one can see that the importance of a discovery is determined much less by the author's specific achievement than by [the question] whether subsequent investigations prove its range of validity to be narrow or wider or perhaps even all-embracing". Boveri's gynander studies, although pioneering in their field, apparently fall in the first of these classes, but his studies on roundworm embryos (of which soon more) doubtless must rank in the latter.

References

Boveri T (1915) Über die Entstehung der Eugsterschen Zwitterbienen. Arch Entwicklungsmech Org [Roux' Arch] 41:264–311

Milne CP Jr (1986) Cytology and cytogenetics. In: Rinderer TE (ed) Bee genetics and breeding. Academic Press, Orlando

Morgan T (1905) An alternative interpretation of the origin of gynandromorphous insects. Science 21:632–634

Morgan T (1907) The cause of gynandromorphism in insects. Am Naturalist 41:715–718

Morgan T (1916) The Eugster gynandromorph bees. Am Nat 50:39–45

von Siebold T (1864) Ueber Zwitterbienen. Z Wiss Zool 14:73–80

Woyke J (1963) Drone larvae from fertilized eggs of the honeybee. J Apicult Res 2:19–24

All translations by the present author, with additions and omissions marked by square brackets or dots. Figures A and B are from Morgan (1905); the hatched nuclei or portions thereof are of paternal origin. The other figures are taken from Boveri's beautiful colour plates of 1915. The half-tone reproductions in this and the previous Essays on Theodor Boveri were skilfully prepared by Sabine Collatz.

ESSAY

Landmarks in Developmental Biology (19)

Of gradients and genes: Developmental concepts of Theodor Boveri and his students

Klaus Sander

Gradients, for decades a "dirty word" in developmental biology (Lawrence 1992), have become respectable of late and are likely to stay with us after their role in pattern formation has been demonstrated at the level of gene regulation (Driever and Nüsslein-Volhard 1988). The present essay is to illuminate the often misunderstood "relativistic" gradient concepts of Theodor Boveri (who actually introduced the term *Gefälle* = gradient), and their influences on his students. Boveri also paved the way, by his studies on chromosomes and development (see Essays 15 and 16), for a chromosomal location of the Mendelian hereditary factors (since 1909 called genes). His most influential students reacted quite differently to his insights on both gradients and genes. Hans Spemann (1869–1941) was remarkably reluctant in linking either concept with his organizing center (see Sander 1986; Wolpert 1986). Waldemar Schleip (1879–1948), however, extended Boveri's idea of interplay between cytoplasmic factors and genes in development, yet maintained a specific "Boverian" definition of gradient. Leopold von Ubisch (1885–1965) finally achieved a synthesis linking the emerging modern concept of morphogenetic gradients with differential gene activity in order to generate spatial patterns.

The hypothetical *Gefälle* that Boveri suggested for the sea urchin egg (see Essay 14) was of the "sequential" type, as defined by Frankel (1989), that is, its function was merely to localize a region of "first differentiation" from where all other regions were to be determined without further recourse to the gradient. This concept, like formally similar ideas of T.H. Morgan and C.M. Child, is quite distinct from the familiar morphogenetic gradient generating the famed French flag pattern via staggered thresholds (see Wolpert 1986). This may in part explain why Wolpert (p. 350) failed to recognize the gradient behind Boveri's sea-urchin concept. Terminology may have been an additional obstacle: it is true that Boveri in his relevant sea-urchin paper (see Essay 14) does not expressly mention gradients. However, failure to use the term does not prove lack of the concept. This is nicely documented, for instance, by Morgan's relevant textbook of 1928: although Morgan had maintained since 1895 that graded properties play a role in regeneration, the entry "gradient" in his register refers only to some principle causing spermatozoa to aggregate!

The term *Gefälle* was used by Boveri expressly in his work on early embryogenesis in *Ascaris* (now *Parascaris*), the horse roundworm. There a gradient (of sorts, see below) was to account for a process discovered by Boveri much earlier in his career, namely, differential chromatin diminution: after first cleavage, the animal blastomere or (mostly) its daughters (see Seidl et al. 1988) selectively show diminution, while its vegetal sister maintains the large germline chromosomes (see figures 26 to 28 on this page). Why should

Fig.26

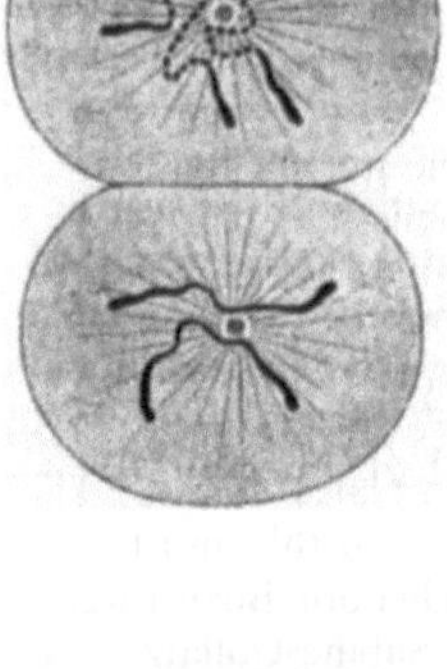

Fig.27

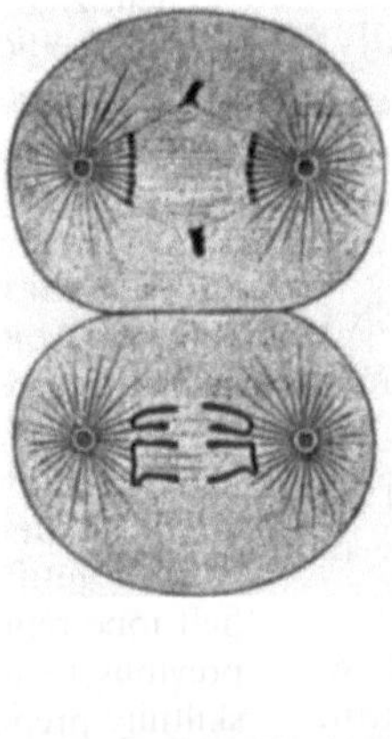

Fig.28

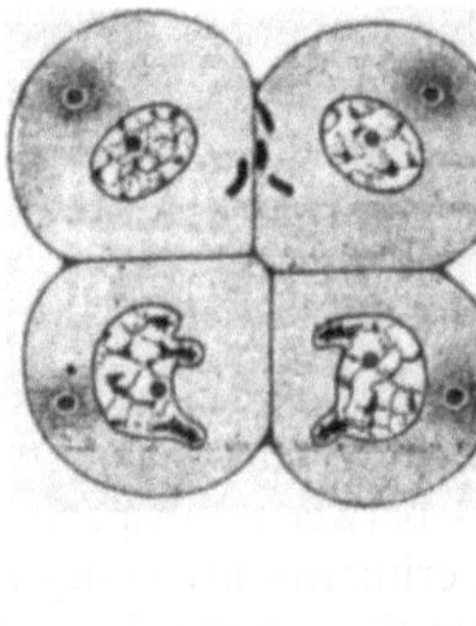

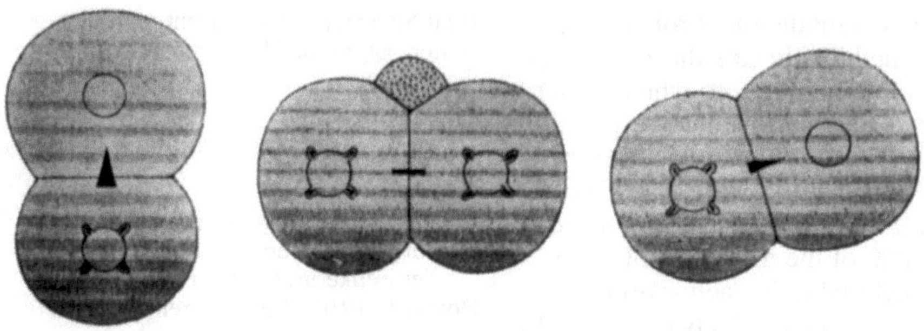

this be so? Boveri (1910) offered two explanations envisioning regional differences in the egg cytoplasm, assumed to be either absolute (i.e. based on locally restricted determinants) or relative.

Boveri's figures illustrating the "relativity hypothesis" are shown on this page. To the modern cursory reader, they pose a treacherous pitfall (first pointed out to me by Karl Moritz). Boveri suggests that the cytoplasm of the egg cell is layered "in such a way that something increases or decreases in concentration from the animal towards the vegetal pole" (as shown for the two-cell stage in Fig. a). Nowadays, and especially after looking at that figure, one would call this layering a gradient – but *not so* Boveri. Instead he reserved the term *Gefälle* for any difference, between the daughter cells, in the distribution (or perhaps total amount) of the "something" they had received from the egg cell. For normal cleavage, where this difference assumedly is large (see Fig. a), Boveri symbolized his gradient by a rather broad wedge extending across the joint cell borders. Where the assumed difference is small or nil (owing to experimental interference), the wedge was made narrow (c) or transformed into a rectangle (b). The point is that even with the slightest difference between the cells, their chromosomal fates will differ as drastically as between the normal blastomeres that show a much larger difference.

To wit: the gradient of stripe width in the cells of Fig. a, b and c merely served to endow one daughter cell with more and the other with less of the hypothetical "something," and it is this difference across the cell borders (named the *Gefälle*) that triggers the ensuing difference in chromosomal fate[1].

As for the chromatin (and implicitly genes) that would direct further development of these cells, Boveri (1910 p 191) drew a much quoted conclusion: "Thus, to me the case of *Ascaris* appears to provide a plain paradigm of how to envision the interaction of protoplasm and nucleus in ontogenesis, and how the extremely small differences in the egg protoplasm may – by their releasing effects on the nucleus, followed by repercussions from the nucleus on the protoplasm – give rise to the ultimately immense differences between the ensuing cells." Boveri's friend E.B. Wilson (1925, p. 1116) used this concept in order to resolve the old enigma of preformation versus epigenesis: "Fundamentally [...] we reach the conclusion that in respect to a great number of characters heredity is effected by the transmission of a nuclear preformation which in the course of development finds expression in a process of cytoplasmic epigenesis."

Waldemar Schleip in 1927 (p. 75) set forth these concepts in more detail: "Genetical research and developmental physiology have shown that whatever can potentially arise is strictly preformed in the genotype, whereas all that really does arise – from the first differentiations in the ooplasm up till the properties of the mature body – results from the genes' reaction to [cytoplasmic] developmental conditions, and thus is generated epigenetically. If the very likely view is confirmed that all genes belong to the nucleus, then one has to accept that the invisibly structured ooplasm results from the first and fundamental reaction [of the genes]. Thereby the genes are brought under new developmental conditions and under these extend their effects. By stepwise increasing intertwinings, this leads to further differentiation within the [former] ooplasm, whereby each level [of differentiation] reached is at the same time the determining cause for new gene activities and thus for the next level. Each achieved state of development determines not only the kind but also the localization of the ensuing differentiations, and since localization, given proper conditions, is leading to a complete, integrated and harmonious organization, individuality as such is something generated epigenetically". Thus, several years before Morgan (1935) unfolded a similar "pleasing picture of the developmental process," Schleip had adapted Boveri's seminal paradigm to progress meanwhile achieved in genetics.

Nor was Boveri's *Ascaris* gradient lost on Schleip. He applied it, for instance, to a result that some 40 years later came to be linked to the "Nieuwkoop Center." Repeating Oscar Schultze's famous *Umkehrversuch* with frog eggs, Schleip and his student Andreas Penners (1890–1951) had noted that the ectopic confrontation of white (vegetal) and dark (animal) cytoplasm can trigger additional embryonic axes. To quote Penners (1929, p. 101): "This generation of a new organizing center apparently is brought forth merely by the influence of white yolk (of course including the affiliated plasmatic substances) on dark yolk." Similarly, but with reference to gradients, Schleip wrote in his *magnum opus*: "There might exist a gradient between cells which are loaded with white yolk and cells which are not, and this might represent the releasing cause" (Schleip 1929, p. 705).

It should be realized that Schleip's was still a "Boverian" gradient across the borders of juxtaposed cells. The "staggered threshold" – or morphogenetic – gradient came into its own only during the next decade, essentially by the work of Runnström and his school (including Hörstadius) in sea urchins, and of Dalcq and Pasteels in amphibians (see Wolpert 1986).

Their concepts provided Leopold von Ubisch with the basis for a decisive step forward – but his achievement did not meet with the deserved recognition, owing to his involuntary retreat to rural Norway (enforced on him by Nazi pressure), and to post-war lack of communication in the scientific community. The essential message of his book of 1953 and an earlier essay was that gradients generate patterns via differential gene activation. A very illustrative comparison (p. 166), to which the present writer has added the figure on this page, starts by quoting Hans Spemann who had problems with Child's gradient concept: "To me it seems that a gradient can only exert an effect if something is really flowing, for instance a stream of water or an electrical current. Even the steepest hillside cannot, as such, drive a water mill" (Spemann 1936 p. 214; the quotation is missing in the English edition of 1938!).

To this, v. Ubisch countered by bringing in the genes: "This argument, impressive as it may seem at first glance, loses significance when we place the gradients in relation to the morphogenetic role of the genes, which is the basic idea of this book. To abide by Spemann's nice imagery: How does it come that at different levels of elevation different floras

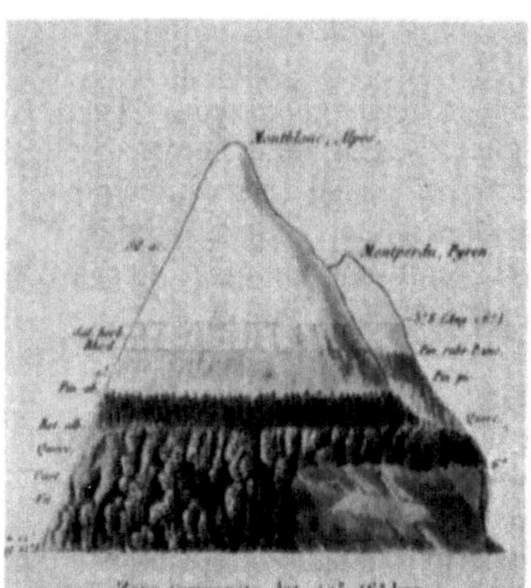

thrive on a mountain? Probably because the hillside cuts through a gradient of temperature to whose different levels are assigned corresponding plants. This correlation can also be expressed the following way. The climatic gradient of the mountain exerts selection: of the multitude of plant seeds offered to it, each level in the climatic gradient selects those that fit it. Exactly this happens in the embryo: the cytoplasmic gradient at each of its levels selects – from the total genome at its disposal in all nuclei – those genes which are correlated (to that level). There is no need for anything to "flow."

It was this concept of von Ubisch that some years later encouraged the writer (Sander 1960, p. 703) to suggest some sort of threshold-dependent gene activation as the link between morphogenetic gradients and the visible segment pattern in insect embryogenesis; which suggestion was printed in small type because (see the opening sentence of this Essay!) gradients were suspect to his *Doktorvater* – all the more so with genes thrown in to complicate matters.

[1] A related, quite important but usually neglected finding of Boveri (see Seidl et al. 1988; Moritz 1993) cannot be discussed here. It concerns the further developmental fate of the equal blastomeres shown in Fig. b, which both behave like vegetal daughter cells rather than egg cells, although they comprise all the cytoplasmic layers of the latter.

All translations are by the present author; square brackets mark omissions or additions. The *Ascaris* figures are from Boveri (1904 and 1910); the figure on the front page is a schematic representation intended "to visualize chromatin diminution with as few figures as possible." The zonation of plants on Montblanc was taken from Alexander von Humboldt's *Prolegomena* of 1815. The author wishes to thank Karl Moritz (München) for helpful advice concerning interpretations of *Ascaris* development, and Sabine Collatz (Freiburg) for reproducing the illustrations. He would also like to express his debt to those (partly unknown) members of his institution who through more than half a century and despite many hardships, including war-time destruction of the building, have preserved Hans Spemann's magnificent reprint collection, an invaluable source in writing these essays.

References

Boveri T (1904) Ergebnisse über die Konstitution der chromatischen Substanz des Zellkerns. Gustav Fischer, Jena

Boveri T (1910) Die Potenzen der *Ascaris*-Blastomeren bei abgeänderter Furchung, zugleich ein Beitrag zur Frage qualitativ-ungleicher Chromosomen-Teilung. Festschrift für Richard Hertwig, vol 3. Gustav Fischer, Jena

Driever W, Nüsslein-Volhard C (1988) The *bicoid* protein determines position in the *Drosophila* embryo in a concentration-dependent manner. Cell 54: 95–104

Frankel J (1989) Pattern formation. Ciliate studies and models. Oxford University Press, New York

Lawrence PA (1992) The making of a fly. The genetics of animal design. Blackwell Scientific Publications, London

Morgan TH (1928) Experimental embryology. Columbia University Press, New York

Morgan TH (1935) The relation of genetics to physiology and medicine (Nobel lecture). Norsted and Söner, Stockholm, p. 12 of reprint

Moritz K (1993) Theodor Boveri (1862–1915), Pionier der modernen Zell- und Entwicklungsbiologie. Gustav Fischer, Stuttgart

Penners A (1929) Schultzescher Umdrehungsversuch an ungefurchten Froscheiern. Roux's Arch Dev Biol 116: 53–103

Sander K (1960) Analyse des ooplasmatischen Reaktionssystems von *Euscelis plebejus* Fall. (Cicadina) II. Roux's Arch Dev Biol 151: 660–707

Sander K (1986) The role of genes in ontogenesis. In: Horder TJ et al. (eds) A history of embryology. Cambridge University Press, Cambridge, pp 363–395

Schleip W (1927) Entwicklungsmechanik und Vererbung bei Tieren. In: Baur E, Hartmann M (eds) Handbuch Vererbungswiss. Bornträger, Berlin

Schleip W (1929) Die Determination der Primitiventwicklung. Akad Verlagsgesellschaft, Leipzig

Seidl C, Bauer M, Moritz KB (1988) Chromatin diminution and early cleavage in *Parascaris univalens*. Roux's Arch Dev Biol 197: 307–320

von Ubisch L (1953) Entwicklungsprobleme. Gustav Fischer, Jena

Wilson EB (1925, reprinted 1928) The cell in development and heredity (3rd edn). Macmillan, New York

Wolpert L (1986) Gradients, position and pattern: a history. In: Horder TJ et al. (eds) A history of embryology. Cambridge University Press, Cambridge, pp 347–362

ESSAY

Landmarks in Developmental Biology (20)

Yves Delage (1854–1920) and the ideology behind his research on *fécondation*

Jean-Louis Fischer

Remembering Yves Delage means reflecting on the relations between science and ideology and retracing a period when experimental embryology put biology in a different light. By his scientific work and the results of his experimental studies, by his efforts to spread biological knowledge and by his teaching, Delage contributed to scientific progress and to establishing the new biology of the years 1890–1910. In embryology, Delage became prominent by his work on merogony and parthenogenesis, which was designed to provide insights into *fécondation*,[1] the processes initiating individual life.

Born in Avignon in the south of France on 13 May 1854, Delage lived there until 1873, the year when he went to Paris to study medicine, submitting (in 1880) his doctoral thesis "On the origins of the formed elements of the blood in vertebrates". At the Sorbonne he attended the courses given by the famous zoologist Henri Lacaze-Duthiers (1821–1901), founder of the marine stations at Roscoff and Banyuls-sur-Mer. He fascinated Delage, who changed course with the new aim of a career in biology and embarked on his first studies in marine biology. In 1881 Delage submitted his thesis for the doctorate in natural sciences, entitled "Contributions to the study of the circulatory organs in édriophthalmic crustaceans". He was appointed to the Sorbonne as *maître de conférences de zoologie* in 1882, subsequently becoming *chargé de cours* in 1885 and, in 1886, *professeur titulaire* (full professor) of zoology in the faculty of sciences in Paris. In 1901 he succeeded Lacaze-Duthier as the director of the marine biology station at Roscoff, which he renamed *Station biologique*. When his eyesight began to fade, he gave up his assigment to the faculty in 1908 and terminated his research in 1912. He maintained the directorship at Roscoff, however, guiding research there and developing a passion for mechanics (for instance in 1912 he introduced the *batyrhéomètre*, an instrument for measuring currents in the sea). He died on 7 October 1920, at Sceaux near Paris (for his life and work, see Goldsmith 1920/21 and Fischer 1979).

The scientific *oeuvre* of Delage is characterized by two important strands of activity: he was an experimenter striving to resolve or understand biological problems with the aid of experimentation, and he was a theoretician and promulgator of the life sciences. That second task played an important part in Delage's intellectual activities. From the beginning of his teaching he felt the need to spread scientific knowledge, and he used two methods for this: treatment of a subject from all angles between the covers of a book, and periodical reviews. A grand synthesis is provided in his book "*La structure du protoplasma, les théories de l'hérédité et les grands problèmes de la biologie générale*" (1895), which in the second edition had its title changed to "*L'hérédité et les grands problèmes de biologie*." Further syntheses were written and published in book form together with Marie Goldsmith: *Les théories de l'évolution* (1909/1914) and *La parthénogenèse naturelle et expérimentale* (1913). In these books Delage took account of current knowledge, analyzing the known facts as well as the existing theories, and then put forward his critique and developed his own hypotheses and theories. For reviews of recent pro-

[1] It is tempting to render the term *fécondation* as "fertilization" in English, but this might be misleading to the modern reader, who is likely to view the union of pronuclei (and hence genomes) as the decisive event in fertilization, while the egg can be activated for embryogenesis by other means as well (as is evident from parthenogenesis). In Delage's time, however, it was widely held that the essential aspect of syngamy is the initiation of development – witness e.g. Wilson's quotation (1896, p. 140) that "the [paternal] centrosome alone causes division of the egg, and is therefore the fertilizating element proper." This is how *fécondation* is to be understood throughout this Essay

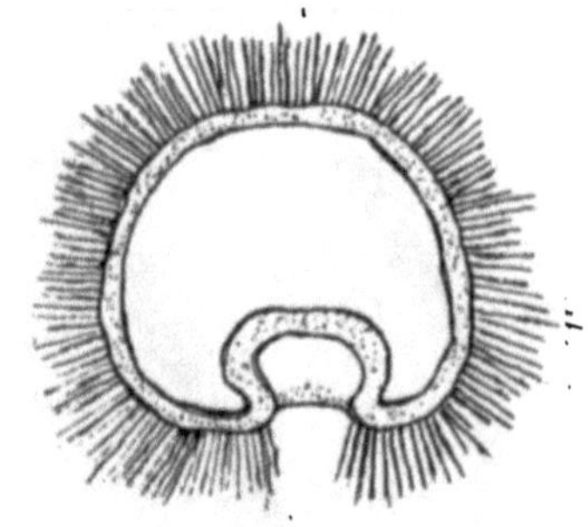

Fig. 4.
Echinus sp.
Larve hémigonique au stade blastula, commençant à s'invaginer (3e jour).

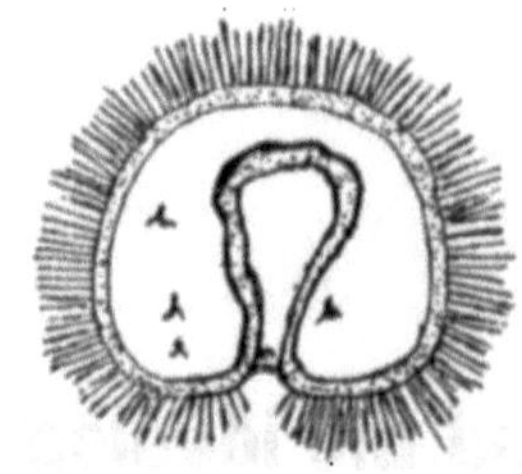

Fig. 5.
Echinus sp.
La larve précédente, plus avancée, ayant complété son invagination (4e jour).

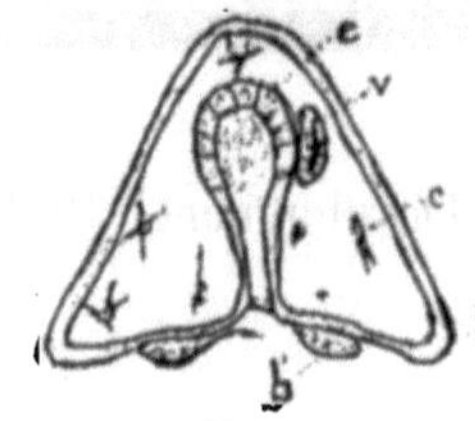

Fig. 6.
Echinus sp.
Pluteus hémigonique provenant de l'évolution de la blastula de la figure précédente (5e jour).

gress in biology, he founded the periodical *L'Année biologique*, which informed biologists and medical people at large on current work related to the great problems of biology – the cell, embryology, heredity, evolution, physiology, etc.

In his comprehensive writings Delage explained the biological facts and also the relevant theories proposed to explain phenomena revealed to the observer and experimenter. Since life cannot be fully grasped, the theoretician of life sciences cannot found his concepts solely on the established facts: unavoidably, there is some ideological component when theories are worked out. The great theoretical debates of the late nineteenth and early twentieth centuries, which arose essentially out of the birth of experimental embryology in the 1880s, exemplify this relation between science and ideology. For instance, a specialist in life sciences could not defend neo-epigenesis against neo-preformation, or neo-lamarckism against neo-darwinism, simply on the basis of experimental results, because the very same results, e.g. those of Chabry on ascidian embryos (Fischer 1992), can be interpreted in opposing theoretical frameworks, in this case neo-preformation and neo-epigenesis.

Delage was neo-Lamarckian and neo-epigeneticist: neo-Lamarckian in accepting that the milieu plays an important part in the mechanism of evolutionary transformation; and neo-epigeneticist in defending the principle of Oscar Hertwig, who had maintained that "in the gastrula, it is not the endoderm which invaginates, but that which invaginates gives rise to the endoderm" (Delage and Goldsmith 1914, p. 157).

Schematically, for the neo-epigeneticist the differentiation of embryonic cells (which in their primary organization are indifferent and undetermined) is a *fonction du lieu*, i.e. depends on the location to which the morphogenetic movements have carried the cells. For the neo-preformationist, however, the undifferentiated embryonic cell is programmed to differentiate into endodermal cells (or cells of other categories and organs). Here the ideological guesswork required for a general concept of life phenomena expresses itself in the choice between freedom (epigenesis) and constraint (preformation), or in the reconciliation of these two terms by invoking one or the other to a smaller or larger degree – as did Albert Dalcq, who wrote in 1953 that "development as a whole appears like a vast epigenesis, but its comprehension requires nonetheless a minimum of preformation." This compromise was rejected by Paul Brien in 1974 (p. 150), who stressed the more categorical alternative that "the living being, by building itself, does not execute a program pre-established and prescient, inscribed in the germ in whatever form. There is no preformism. The living makes itself in the making, by epigenesis."

On this basis, Delage also rejects Darwin's theory of pangenesis and other similar theories that represent a certain prefiguration of Johannson's gene concept of 1910 (see Fischer and Schneider 1990). He refused to accept these "micromeristic" theories, which were put forward by H. Spencer (polarigenesis; 1864), E. Haeckel (perigenesis of the plastidules, 1876), R. Altmann (granules, bioblasts; 1980/1894), K. Nägeli (micelles, idioplasm, 1884), H. de Vries (intracellular pangenesis; 1889), A. Weismann (determinants; 1892) etc. By contrast, Delage favoured some kind of *organicisme*[2], to the degree that he considered it a *conception fondamentale* and the *plus juste* and *plus féconde*. It was Wilhelm Roux (Sander 1991) who, in 1881, published a partial elaboration of this theory embodying the concepts of functional stimulation, of self-differentiation of cells and organs and of self-determination of functions. The individual and the species are products of a sequence of causal chains of biological events, but these cannot be reduced to "factors" that between them predetermine the origins of the individual and the species.

The micromeristic theories of Weismann or de Vries had placed such determining factors in the cell

[2] Delage defined this therm as the "concourse of some moderate determination and some permanently acting ambient forces, which are always necessary not as a simple precondition but as an essential element of the ultimate determination of development" (Delage 1903, p 750), but he himself rejected any kind of preformation

nucleus, and more specifically in the chromosomes. In order to demonstrate that, on the contrary, the cytoplasm plays an important biological role, Delage (1898, 1899a, b) put forward a method, called *mérogonie*, of curtailing the role of the maternal pronucleus and evaluating the role of the "*protoplasme*" in biological development. Merogony ("generation from a part") ensues from fertilizing a fragment of enucleated cytoplasm cut from an egg cell. Delage developed this technique in a precise fashion and extended it to various organisms (echinoderms, molluscs and annelids), following several earlier attempts by Boveri (see Boveri's historical note of 1901).

The idea of fertilizing chunks of ooplasm lacking the maternal pronucleus was originally conceived by the Hertwig brothers in 1887. They shook sea urchin eggs violently in a test tube for half an hour and obtained "drops" (or fragments) of cytoplasm. However, their fragments did not show any development even when provided with spermatozoa, although the sperm that had entered the fragments formed mitotic spindles there. Using the same technique in an interspecific experiment, Boveri in 1889 obtained larvae of typically paternal appearance (see Moritz 1993). But since Boveri had produced his fragments by the shaking method, his results were subject to doubts, especially those from the first paper where he had not ascertained the absence of the nucleus under the microscope. It was only by using a direct technique, i.e. cutting the ooplasm and isolating the enucleated fragments under the microscope, that Delage could confirm Boveri's result. He obtained some merogonic larvae from sea urchin, scaphopod and annelid species (see the figures). This strengthened his view (Delage 1899, p. 648) that *fécondation* results essentially "from the union of a sperm nucleus with a mass of egg cytoplasm, and that this confers on the cytoplasm some special energetic plasm contained in the spermocentre."

After chemical parthenogenesis had been achieved in the sea urchin by Jacques Loeb in 1899/1900, numerous embryologists jumped on this new line of research. Delage was among them and from 1900 to 1913 published 33 notes and papers on this topic, and also a book (Delage and Goldsmith 1913). Together with Loeb, he became one of the leading practitioners and theoreticians of animal parthenogenesis. At the 8th International Congress of Zoologists in Graz (15–20 August 1910), Delage presented a synthesis of work on experimental parthenogenesis and concluded that the effects of all chemical substances triggering parthenogenesis (salts, acids, bases), carbon dioxide and physical treatments (warmth, electricity) could be reduced to a common factor, a modification of the ooplasm that would enable it to enter on parthenogenetic development.

With his research on experimental parthenogenesis, Delage was the first (in 1901) to demonstrate the fundamental role of ooplasmic maturation for the onset of development. To him, his studies on experimental parthenogenesis were a logical continuation of those on merogony, because both experiments guided the egg cell (or a part of it) into embryogenesis despite the absence of any union of pronuclei (amphimixis). Moreover, since merogony excluded the maternal pronucleus from initiating development and parthenogenesis excluded the paternal pronucleus, the ooplasm was left as the crucial component in triggering development – a concept that was all the more attractive from the neo-epigenetic and neo-Lamarckian point of view, because the ooplasm would be more receptive to environmental stimuli than the nuclei.

These same experiments, however, highlighted the problem of the role the nucleus might play in *fécondation* and, more generally, in the formation of the individual. The

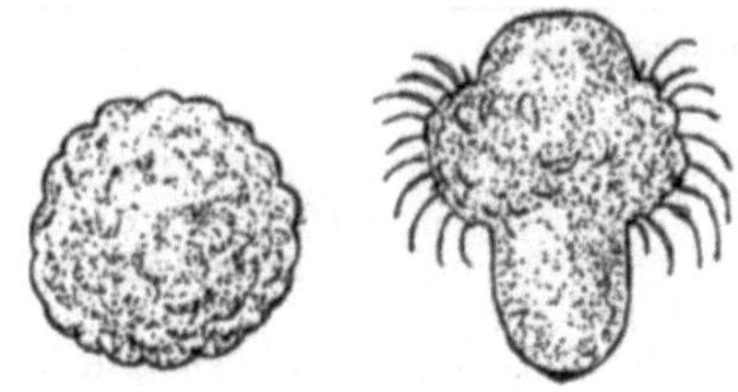

Fig. 7.
Dentalium entale.
Embryon et larve hémigoniques.

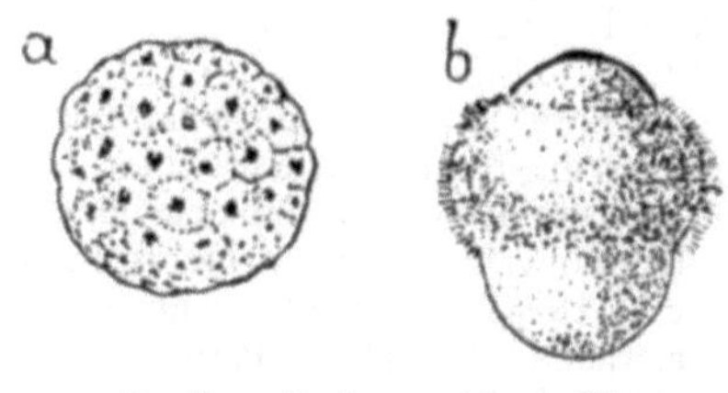

Fig. 30. — Embryon et larve d'Annélide (*Lanice*) provenant de la fécondation mérogonique.

experiments showed that development could occur without either the paternal or the maternal nucleus being present. Consequently, Delage and the majority of neo-Lamarckians and neo-epigeneticists refused to ascribe a fundamental role to the *unités morphologiques* (the chromosomes or particles constituting the chromosomes) and could not bring themselves to reduce the destiny of the future individual to the activities of those units. They preferred to think that biological events are the responsibility of physicochemical rather than morphological structures. In this they were in fact not wrong; but one must reproach them for having rejected the structural level of the chromosomes, which later so efficiently was to provide access to the chemical level of development.

Thus, in the general history of developmental biology, the period of merogony and chemical parthenogenesis is important to our understanding of certain choices of theoretical positions, whether by Delage or by biology in general. It ended in 1910/1911 in an epistemological break, triggered through the achievement of traumatic parthenogenesis by Eugène Bataillon (see the next Essay) which led to the definitive statement that *fécondation* is the result of egg activation plus regulation.

References

Boveri T (1901) Merogonie (Y. Delage) and Ephebogenesis (B. Rawitz), neue Namen für eine alte Sache. Anat Anz 19:156–172

Brien P (1974) Propos d'un zoologiste le vivant epigenèse evolution épigénétique. Editions de l'Université de Bruxelles, Bruxelles

Dalcq A (1953) Préformation et épigenèse dans leur acception actuelle (avec la notion d'epigenèse trophique). Bull Acad R Belg [Classe Sci] 39:1128–1142

Delage Y (1895) La structure du protoplasma, les théories de l'hérédité et les grands problèmes de la biologie générale. Reinwald, Schleicher Frères, Paris

Delage Y (1898) Embryons sans noyau maternel. C R Acad Sci (Paris) 127:528–531

Delage Y (1899a) Sur la fécondation mérogonique et ses résultats. C R Acad Sci (Paris) 129:645–648

Delage Y (1899b) Etudes sur la mérogonie. Arch Zool Exp Gen 7:383–417

Delage Y (1901) Etudes expérimentales sur la maturation cytoplasmique et sur la parthénogenèse artificielle chez les échinodermes. Arch Zool Exp Gen 9:285–326

Delage Y (1903) L'hérédité et les grands problèmes de la biologie générale. Reinwald, Paris

Delage Y, Goldsmith M (1913) La parthénogenèse naturelle et expérimentale. Flammarion, Paris

Delage Y, Goldsmith M (1914) Les théories de l'évolution, 2nd edn. Flammarion, Paris

Fischer JL (1979) Yves Delage (1854–1920): l'épigenèse néo-lamarckienne contre la prédétermination weismanienne. Rev Synthèse 95–96:443–461

Fischer JL (1992) The embryological oeuvre of Laurent Chabry. Roux's Arch Dev Biol 201:125–127

Fischer JL, Schneider HW (1990) Histoire de la génétique. A.R.P.E.M, Paris

Goldsmith M (1920/1921) Yves Delage (notice nécrologique). Ann Biol XXV:V–XIX

Moritz KB (1993) Theodor Boveri (1862–1915), Pionier der modernen Zell- und Entwicklungsbiologie. Fischer, Stuttgart

Sander K (1991) Wilhem Roux and his programme for developmental biology. Roux's Arch Dev Biol 200:1–6

Wilson EB (1896) The cell in development and inheritance. Macmillan, New York

The portrait, obtained through the kind services of professor J.C. Beetschen at Toulouse, shows Delage working in his attic laboratory at Roscoff, probably around 1900. The drawings (from Delage 1899b) show merogonic embryos of the sea urchin, an elephant tusk snail (*Dentalium*) and an annelid worm (*Lanice*).

Translation of the French manuscript by K. Sander, Freiburg.

Landmarks in Developmental Biology (21)

Eugene Bataillon (1864–1953) and traumatic parthenogenesis

Jean-Louis Fischer

It is to Eugene Bataillon (Courrier 1954; Rostand 1959; Fischer and Smith 1984) that we owe the insight that the *fécondation* of the ovum results from two events, first the *activation* by the penetrating sperm, and second a *régulation* induced by the sperm aster, which initiates a balanced cell division, the first step in embryogenesis.[1] Bataillon put forward this hypothesis after his discovery (1910/11) of traumatic parthenogenesis in a vertebrate, the frog.

Like Delage (see the previous Essay) and numerous other embryologists, Bataillon (1955, in his "scientific testament") vividly recalls the interest awakened in him by Jacques Loeb's initial results on experimental parthenogenesis. In 1899 Loeb had obtained parthenogenetic sea urchin larvae after exposing the eggs to a hypertonic solution of $MgCl_2$ (see Delage 1912 for a detailed bibliography of early work on experimental parthenogenesis).

Loeb believed that the magnesium ion exerted a specific action in triggering parthenogenetic development. In the case of normal fertilization the spermatozoon would carry the ions responsible for the mechanism promoting *fécondation* and the cytological events leading to ontogenesis.

Bataillon was also influenced by the work of N.M. Kulagin (1898), who, having submerged the eggs of fish and amphibia in antidiphtheria serum, had observed furrowing as in cleavage stages, which he ascribed to an excitation akin to that normally conferred by the spermatozoa.

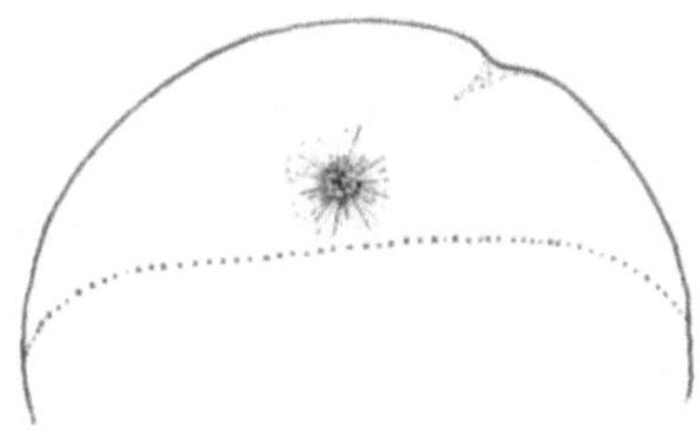

Fig. 8.

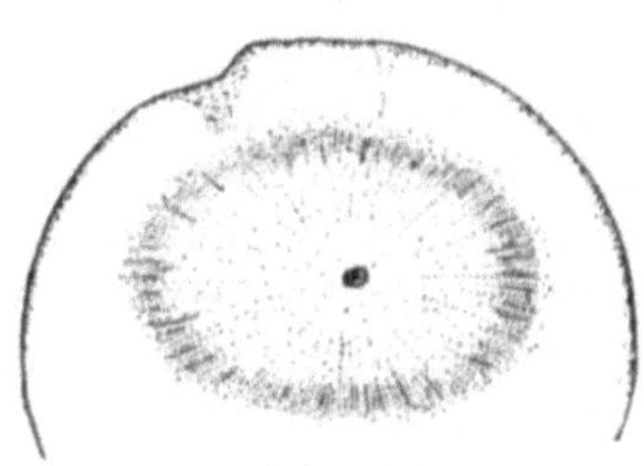

Fig. 10.

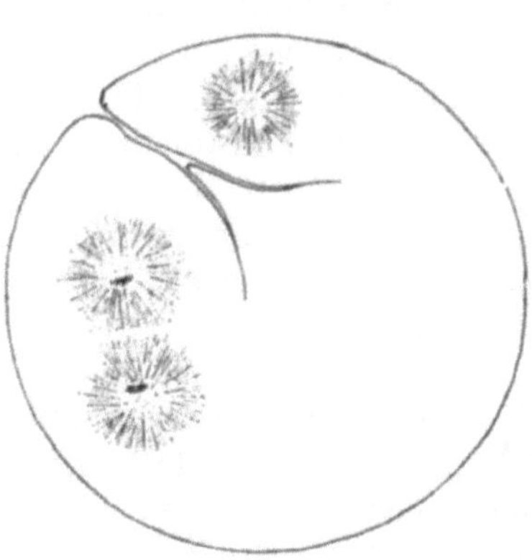

Fig. 13.

These observations on experimental parthenogenesis led Bataillon to postulate some effect of osmotic pressure as the agent triggering embryonic development (Bataillon 1901). He therefore performed the following experiment. Unfertilized eggs of fish or amphibians were divided into three lots to be placed in the following solutions: (1) serum from cattle blood (with an osmolarity corresponding to 0.9% NaCl), (2) 0.9% NaCl, and (3) sugar (saccharose) at the same osmolarity. After one or two hours the eggs were transferred to plain water. All three batches reacted in the same way: they underwent several cleavage divisions, preceded (in the case of amphibians) by rotation. However, they developed no further and thus parthenogenesis was abortive. These results Bataillon published in the *Comptes rendus de l'Academie des Sciences de Paris* on July 9, 1900.

In August 1900 Loeb achieved parthenogenetic development in the

[1] *Activation* will release the egg cell from its quiescent state but *per* se this would be followed by monaster formation and aberrant cleavage (see figures marked 10 and 13). A second process, *régulation*, is therefore needed for spindle formation and cleavage to take their normal course; this step thus *"corrects"* the insufficiency of a mere activation.

sea urchin after exposing the eggs to hypertonic solutions of non-electrolytes (saccharose, urea). Loeb adopted the same views as Bataillon to a degree, but he thought that the surrounding milieu, owing to its hypertonicity, caused some loss of water from the unfertilized egg; according to this osmotic pressure would play its part in parthenogenetic development by modifying the chemical composition of the egg cell. Both Loeb and Bataillon obtained, however, only the first steps of development in fish and amphibian species by raising osmolarity and in the sea urchin the best result was the achievement of larvae described as *fragiles, rampantes* and *peu viables.*

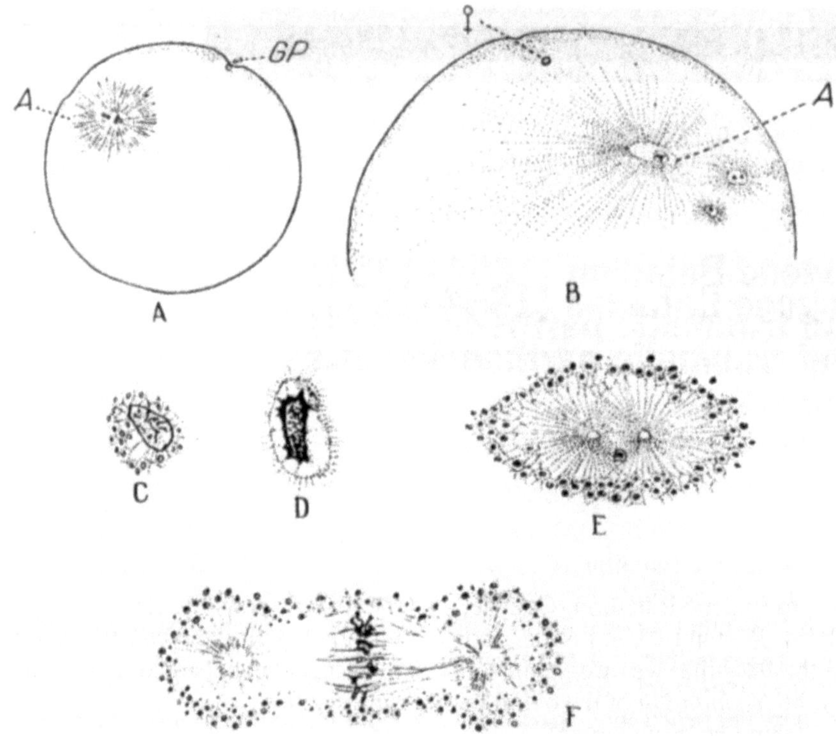

Earlier, Oscar and Richard Hertwig had exposed sea urchin eggs to chloroform and observed the raising of a fertilization membrane (or vitelline membrane; Bataillon preferred to call this membrane the *membrane d'activation* when it was formed during parthenogenesis). The Hertwigs concluded that this treatment had a stimulating effect analogous to that of the spermatozoon (see Delage 1912). In 1893 Curt Herbst repeated the experiment and considered the membrane-raising effect of chloroform to be identical to that of the spermatozoon.

These experiments conducted by the Hertwigs and by Herbst inspired Loeb (1905) to develop two-step parthenogenesis. In a first step, he treated sea urchin eggs with a substance raising the vitelline membrane (chloroform, xylene, silver salts, diluted acetic acid) and then, in a second step, exposed them to a hypertonic solution; the sequence could also be reversed and still yield the identical result: first formation of a fertilization membrane, then cleavage divisions. In this way Loeb obtained vigorously swimming pluteus larvae. Loeb considered this treatment an imitation of normal fertilization: the spermatozoon must contain an activating substance (with membrane-raising or cytolytic properties) and a second substance that would correct[1] the insufficient effect of the former, so that normal cleavage could ensue. For Loeb, then, the two steps of experimental parthenogenesis made it possible to reduce normal *fécondation* conceptually to two steps: activation and correction.

Having learned of Loeb's two-step parthenogenesis, Bataillon thought that this method might constructively also be applied to anuran eggs. However, his preliminary experiments yielded nothing but failures. Disheartened, he turned to heterogenic in vitro hybridization in amphibians (anuran × urodelean species) and observed that in certain combinations the mere contact of a urodelean spermatozoon, not even followed by penetration through the cell membrane, was sufficient to activate an anuran egg cell. In 1910 he arrived at the conclusion that, like low or high temperature and ambient hypertonicity, foreign spermatozoa could induce abortive cleavage in fish and amphibian eggs.

When reviewing these findings in his "scientific testament", Bataillon wrote the following sentence, which is instructive for grasping the logic of his thoughts and understanding the scientific work that was yet to follow: "One Sunday in March 1910, I sat hypnotised in front of my microscope, contemplating an impressive sight: a preparation of eggs of the toad (*Bufo calamita*) exposed to sperm from the newt (*Triturus alpestris*). The eggs' surfaces were larded with these alien male elements, whose voluminous heads appeared on the slides like so many surgeon's stylets" (Bataillon 1955, p. 24). It was at this moment that the idea suddenly came to him that "a slight traumatism such as a prick from a sharp glass or metal needle could be as effective as heat or hypertonicity. "

This insight was no doubt related to Chabry's influence. Chabry had taught his techniques to Bataillon during the two years they spent together, which had given Bataillon an aptitude for "tinkering about" (Rostand 1959; Fischer 1992; Sander and Fischer 1992). As soon as the idea had struck him, Bataillon "at once" prepared "glass stylets" and put eggs from a mature female on several watch glasses. The result was

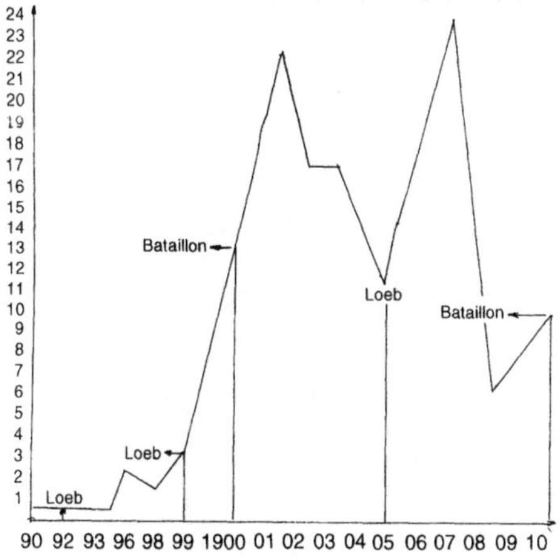

beyond all expectations: he obtained 90% of abortive development, about 10% of rudimentary embryos, but also a few cases of swimming larvae. However, he was afraid that the latter might have derived from eggs contaminated by sperm. He repeated his experiments and, all doubts gone, published his note on 18 April 1910 in the *Comptes Rendus de l'Académie des Sciences de Paris*, and a full paper on November 20 of the same year (Bataillon 1910). At long last he had managed to obtain in vertebrates what Loeb had obtained about five years earlier in invertebrates.

However, the theoretical interpretation was not simple. At the end of 1910, while re-reading the latest volume of the *Année Biologique*, he came across a note by Guyer: "The development of unfertilized frog eggs injected with blood" (*Science* 1907, 25:910–911). Although parthenogenesis had not been Guyer's main preoccupation Bataillon pondered over the problem and saw in his imagination the blood that had contaminated the eggs during his early experiments on traumatic parthenogenesis. Wishing to adapt for the frog the experiment in which Loeb had activated virgin sea-urchin eggs by low concentrations of cyanide, Bataillon (1911) discovered that fairly high concentrations (0.5–1%) of this chemical dissolved the coat that covers the eggs. Quite fortuitously, Bataillon now had a technique at his disposal that was to enable him to carry out the crucial experiments. Eggs from which the coat had been removed were divided into three lots. The first was injected with serum (horse), the second with leucocytes (horse),[2] and third with erythrocytes (horse), each group of eggs being subjected at the same time to the stimulating traumatic effect of the injection. With the serum, no regular cleavage was obtained; with the red blood cells about 1% of regular cleavage was observed; with leucocytes, the proportion of eggs passing cleavage and gastrulation rose to 75%. Bataillon supplemented these results with cytological studies and concluded that experimental parthenogenesis is initiated by two steps: *activation* by the pinprick, and *régulation* by introduction of a nucleus (Bataillon 1911).

Bataillon's greatest merit is that he opened up the way for studies on experimental parthenogenesis in vertebrates and interpreted traumatic parthenogenesis as a "*plagiat fidèle de la fécondation*" (Bataillon 1955, p. 55).

References

Bataillon E (1901) La pression osmotique et les grands problèmes de la biologie. Arch Entwicklungsmech Org [Roux' Arch] 11:149–184

Bataillon E (1910) Le problème de la fécondation circonscrit par l'imprégnation sans amphimixie et la parthénogenèse traumatique. Arch Zool Exp Gén 6:103–135

Bataillon E (1911) Les deux facteurs de la parthénogenèse traumatique chez les Amphibiens. CR Acad Sci Paris 152:920–922

Bataillon E (1912) La parthénogenèse des Amphibiens et la fécondation chimique de Loeb. Ann Sci Nat Zool 16:249–307

Bataillon E (1929) Analyse de la fécondation par la parthénogenèse expérimentale. Roux's Arch Dev Biol 115:707–778

Bataillon E (1955) Une enquête de trente-cinq ans sur la génération 1900–1934. Préface de J. Rostand. SEDES, Paris

Courrier R (1954) Notice sur la vie et les travaux de Eugène Bataillon. Institut 25:1–45

Delage Y (1912) La parthénogenèse expérimentale. Verhandlungen des VIII. Internationalen Zoologen-Kongresses zu Graz. Fischer, Jena, pp 100–162

Fischer JL (1992) Laurent Chabry (1855–1893) and the "Age d'or" of biology. Roux's Arch Dev Biol 201:61–64

Fischer JL, Smith J (1984) French embryology and the "mechanics of development" from 1887 to 1910: L. Chabry, Y. Delage and E. Bataillon. Hist Philos Life Sci 6:25–39

Rostand J (1959) Hommage à Eugène Bataillon. Paper presented at the 84th Congress of the Sociétés Savantes de Paris et des départements à Dijon, Science section

Sander K, Fischer JL (1992) How to dart ascidian blastomeres: the embryological micro-tools of Laurent Chabry (1855–1893). Roux's Arch Dev Biol 201:191–193

[2] Bataillon used horse blood because in mammalians the circulating erythrocytes have lost the nucleus whereas in the frog they retain it. He thus could be sure that the nuclei which acted as initial "regulators" of development (see the figures marked A–F) came from leukocytes.

The figures on the front page render sections through frog eggs activated by pricking with a glass needle, showing monaster formation, aberrant mitosis and irregular cleavage; for technical reasons, these figures are taken from a later paper by Bataillon (1929). The figures on the second page are from frog eggs inoculated with horse leucocytes. Letter A in the top figures marks the aster formed around a leucocyte nucleus; this nucleus is also shown in D and (below the center) in E; C is the maternal pronucleus, F its first mitotic division (from Bataillon 1912). The graph shows the annual output of publications on experimental parthenogenesis from 1890 to 1910, with the main papers of Loeb and Bataillon marked; it is based on the 180 relevant citations in Delage (1912), who in addition listed seven papers published between 1847 and 1888.

Translation of the French parts of the original manuscript by K. Sander.

Landmarks in Developmental Biology (22)

Hilde Mangold (1898–1924) and Spemann's organizer: achievement and tragedy

Peter E. Fässler and Klaus Sander

The organizer effect was the only achievement that the Nobel committee mentioned when awarding the 1935 prize for physiology or medicine to Hans Spemann. The paper by Spemann and Hilde Mangold describing this effect – published in this journal in 1924[1] – had a great impact for some time but then gradually lost in attraction. However, its citation rate rose again – quite dramatically indeed – when some interactions relevant to "Spemann's organizer" became amenable to molecular analysis (Fig. 1).

Spemann's name, by the way, was not added to the term organizer until recently – not so much in remembrance of Spemann, who introduced the term and concept, but rather in order to distinguish it from the "Nieuwkoop organizer", which was subsequently described in this journal (Nieuwkoop 1969). Spemann's life and work were the topic of books written by two of his former students (O. Mangold 1953; Hamburger 1988), and further information can be gathered from his truncated autobiography and the accompanying recollections of friends and colleagues (Spemann 1943) as well as a host of secondary literature (see Fässler 1995).

But who was Hilde Mangold? There is only some rather personal information about her available in print, based on the recollections of her fellow student Viktor Hamburger (1984, 1988). The purpose of the present Essay – appearing in the last issue to bear the title *Roux's Archives*[2] – is to outline Hilde Mangold's life and work, taking advantage of unpublished information and original documents not known of or referred to previously.[3] In addition, answers will be given to some critical questions concerning the validity and originality of Hilde's work and the concepts derived therefrom, and whether her achievements were properly acknowledged.

Curriculum vitae of Hilde Mangold, neé Pröscholdt

Hilde Mangold was born on 20 October 1898, to Gertrud Pröscholdt (neé Blödner) and her husband Ernst, as the second of three daughters. Her father came from a family of artisans specialized in painting decorations on chinaware. He had moved to Gotha from his small home town in Thuringia and started working as a clerk in a soap factory. Gotha was a provincial centre of intellectual life that had come to flourish under its local rulers (linked to British royalty through Queen Victoria's prince con-

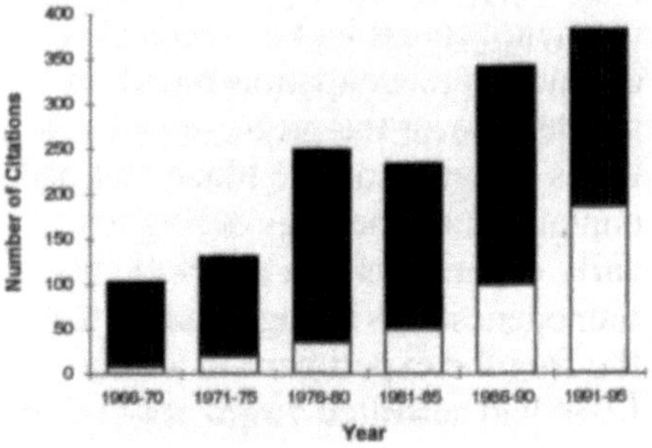

Fig. 1 Number of references made, during successive five-year periods, to publications of Hans Spemann, with the "organizer paper" (Spemann and H. Mangold 1924) accounting for the lower part of each column. Graph based on the ISI® Citation Index; the data for 1995 end with the October issue

[1] From 1923 to 1925, *Roux's Archives* was combined with another journal under the joint title *Archiv für mikroskopische Anatomie und Entwicklungsmechanik*. For a translation of this paper, see Willier and Oppenheimer (1964).

[2] The Journal will continue under the name *Development, Genes and Evolution*.

[3] This material was gathered over the years by K.S. and was recently supplemented by P.E.F., who has analysed it as part of his doctoral thesis (Fässler 1995).

Fig. 2 Marketplace in Gotha with the castle in the background, engraved in mid-19th century. Hilde Pröscholdt's parents owned the prominent building with the stepped gables (*centre*), and it was there that she was born and 23 years later celebrated her marriage to Otto Mangold. Her portrait on the title page is probably from that time. (Courtesy of Rose Lakschewitz, Königsfeld)

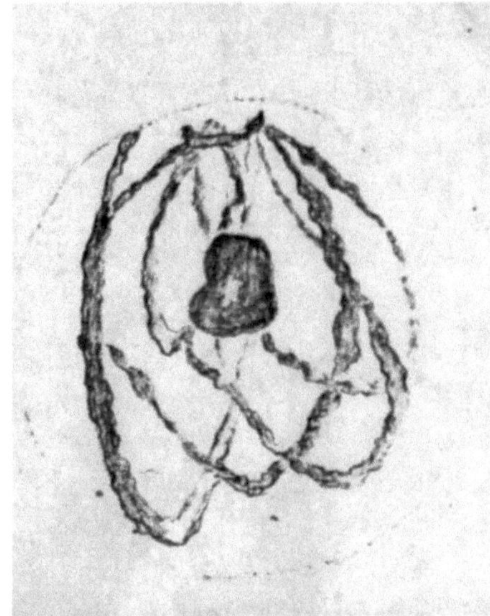

Fig. 3 Pencil sketch from the student's notebook of Hilde Pröscholdt, showing meiotic chromosomes at the "bouquet stage" (see the pun in Fig. 4); this work was supervised by Fritz Baltzer (see Hamburger 1984). (Courtesy of Dr. Otto Burkhardt, Crailsheim)

sort Albert). Proving himself very capable, Ernst Pröscholdt soon moved up through the factory's ranks and in the end married the owner's daughter. She was a gifted person, strongly interested in fine arts and politically engaged in the womens' rights movement.

Hilde's parents were comfortably situated (Fig. 2) and, in keeping with the ideals of the German *Bildungsbürgertum*, provided her with the best education available. Access to this, however, depended not only on money but even more so on intellectual capability. It was on account of the latter that Hilde progressed through various local schools and at age 16 began to attend the famous *Gymnasium Ernestinum* – then almost inaccessible to girls. She passed her final high school exams with flying colours; thereafter, however, she was not sent to university but rather to a private institute introducing young ladies from established families to proper housekeeping and social etiquette. She does not seem to have been enamoured by this, because after half a year, in the winter of 1918/1919, she enrolled at the nearby University of Jena as a student of chemistry; in addition, she attended voluntary courses in philosophy and the history of fine arts. After one semester she moved to Frankfurt/Main, where her main subject was zoology.

It was in Frankfurt that Hilde heard a guest lecture by Hans Spemann. Much impressed – and in keeping with the scholarly tradition of German students – she moved on again, this time to Freiburg, where Spemann had just been called to the chair of zoology. There she attended a broad range of lectures for two semesters and did practical laboratory work, the scope of which was fairly modern for that time (Fig. 3). This work was supervised by Fritz Baltzer, Otto Mangold and Bruno Geinitz. Among her fellow students were Viktor Hamburger and Johannes Holtfreter (Fig. 4).

After 1 year, Hilde asked Spemann for a topic on which to work for her Ph.D. thesis. Spemann apparently was aware of her drive, tenacity and manual dexterity, and accordingly set her a demanding task: she was to turn *Hydra* inside out, as described by Abraham Trembley almost two centuries earlier, in order to see whether ecto- and endoderm would trade characters in keeping with their new positions. Although successful in simpler experiments (Fig. 5), Hilde was unable to repeat Trembley's feat, and Spemann

Fig. 4 Part of a cartoon drawn by Johannes Holtfreter for the party to celebrate the completion of the two-semester practical course. The participants – among them Hilde Pröscholdt (eyes cast down bashfully) and Viktor Hamburger (shaded by vertical hatching) – are shown as the horses pulling Spemann's coach; for a complete reproduction, see pp 224–225 in Horder et al. (1986). The pun involving Hilde and Viktor (top row, third box from the left) could be rendered as "These two are seeking eggs / and bouquets – in long-horned locusts: / Hamburger is trotting to the right of the two / in order not to obscure little Pröscholdt". The bouquet configuration is part of the meiotic prophase (see Fig. 3), but note that *Bukett* also means the decorative bunch of flowers customarily presented to a newly wed couple!

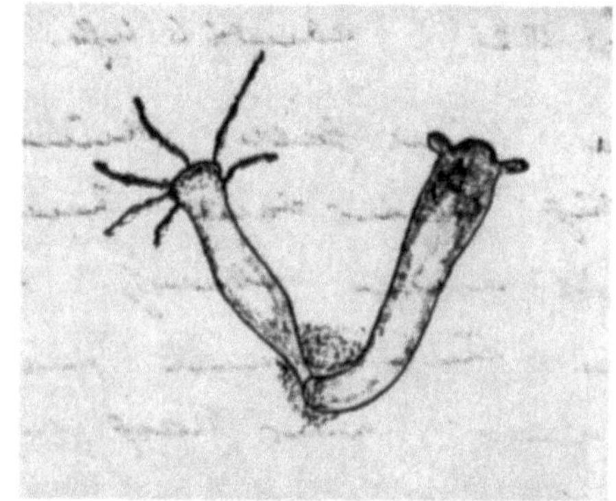

Fig. 5 Sketch drawn by Hilde Pröscholdt to illustrate one of her *Hydra* results, dated 4 March 1921. On 16 February, two individuals (one stained blue) had been cut below the tentacles and the two "heads" were joined together with opposite polarity. Two body columns grew out and finally separated (as shown here). Among the 17 *Hydra* experiments on record, this one came closest to the inductions described by Ethel Browne (see below). (Copy courtesy of the Hubrecht Laboratory, Utrecht)

himself also failed. With valuable time lost in this way – Ph.D. students had to pay considerable fees then! – Spemann apparently felt the need for compensation. Consequently he suggested that she perform an experiment on newt embryos that he knew would technically work: to transplant dorsal blastopore lip material into the prospective belly region of a host embryo from another species. As the respective embryos differed in the amount of melanin granules stored in their animal cap cells, the chimaeras would reliably reveal the respective contributions of graft and host to the outcome of transplantation experiments. Having done all the preparatory work (see below), Spemann must have longed to do this experiment himself, but because he was overburdened with administrative and teaching duties during his first few years at Freiburg, he did not have the time necessary. After carrying out some relatively simple experiments for practice purposes, Hilde successfully performed several hundred of these operations. She submitted her thesis in February 1923 and it was accepted with marks ranging between "best" and "second" (Fig. 6). Soon afterwards she passed the *Rigorosum*, the oral Ph.D. examination, comprising three subjects chosen by herself: zoology, botany and philosophy (where Edmund Husserl, the famous phenomenologist, found her performance "very satisfactory").

In 1921, on her 23rd birthday, Hilde Pröscholdt had married Otto Mangold, Spemann's first Ph.D. student and his "right-hand man" at that time, who was later to be his successor at the Kaiser-Wilhelm-Institut at Berlin-Dahlem and took over his Chair in Freiburg. The mutual attraction of the young couple must have rested on the contrast between them rather than on evident similarities, because Otto came from a rural family – his father being a farmer and innkeeper in a small Swabian village – and there is little indication that he had intellectual ambitions comparable to those of Hilde (for these, see Hamburger 1984). However, they certainly shared a strong interest in the riddles of embryogenesis, and her ability at experimenting plus his steadiness and thoroughness (e.g. O. Mangold 1932) made them a good research team. After their joint work in the spring of 1923 (see H. Man-

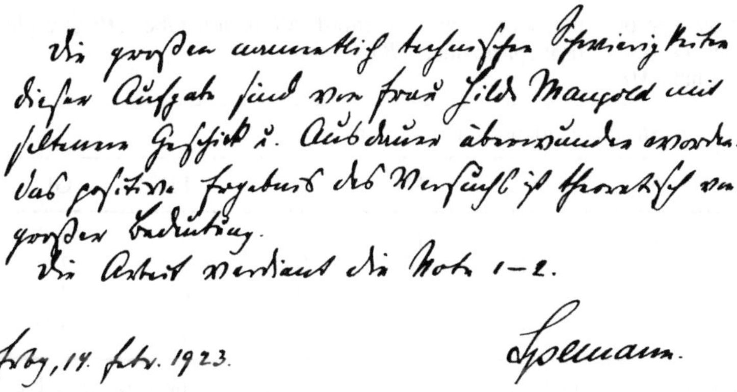

Die großen, namentlich technischen Schwierigkeiten dieser Aufgabe sind von Frau Hilde Mangold mit seltenem Geschick u. Ausdauer überwunden worden. Das positive Ergebnis des Versuchs ist theoretisch von großer Bedeutung.
Die Arbeit verdient die Note 1–2.

Frbg, 14. Febr. 1923. Spemann.

Fig. 6 Closing paragraph of Spemann's report to the faculty on the doctoral thesis of Hilde Mangold. The full text on the single page provided for this purpose reads as follows: "Various experiments had permitted the conclusion that in a young amphibian embryo a limited region is distinguished before the others by determining the development of the indifferent other parts. From this followed the task of testing whether and to what degree small parts of the "organization centre" possess the capacity, when transplanted into foreign indifferent tissue of another embryo, to call forth there the formation of an embryonic rudiment. To allow distinguishing the transplanted "organizer", and what it develops into, from that which it has induced, it was necessary to perform such transplantations between embryos from different species. / The large, especially technical, difficulties of this task were overcome by Mrs. Hilde Mangold with rare dexterity and perseverance. The positive result of the experiment is of great theoretical importance. / Frbg [Freiburg], 14 February 1923 / Spemann". Original in the archives of the Albert-Ludwigs-University, Freiburg i.Br. Reduction 50%

Fig. 7 Hilde Mangold with her son Christian, apparently both dressed up for some public appearance. The photograph was taken in Freiburg, probably early in the spring of 1924. (Courtesy of Rose Lakschewitz, Königsfeld)

Fig. 8 Tombstone of Hilde Mangold at the main cemetery of Gotha. The bronze plaque was modelled by the painter Julius Bissier, a friend of the Mangolds in Freiburg

gold 1929), Hilde's laboratory notes ended – obviously because she was diverted to another task, namely taking care of her son Christian, who was born in December 1923 (Fig. 7). Some months earlier, Otto had received the invitation to take over at Dahlem, and Hilde was delighted, although this meant they had to leave Freiburg (which then, as now, most residents find hard). As she wrote to her sister: "I am terribly pleased for Otto, it is quite an independent position. [...] Otto actually may be a bit too young, and Spemann would have liked to keep him for another couple of years. But it is such a splendid offer that we have no choice."[4]

Less than a year later Hilde was dead. A letter that Spemann wrote to Fritz Baltzer[5] recounts the tragic events. During a visit to Otto's rural home, she was about to refuel the alcohol stove in order to warm

[4] Letter to her sister Gertrud of 10 October 1923, kept in the authors' institute.
[5] Letter of 18 September 1924, kindly made available by the Senckenbergische Bibliothek, Frankfurt/Main.

Table 1 Hilde Mangold's organizer experiments listed by type of chimeric combination and secondary structures observed (Alp *T. alpestris*, Crist *T. cristatus*, Taen *T. taeniatus*; *GL* gut lumen, *ME* mesoderm, *NC* notochord, *NT* neural tube, *OP* otic placode, *PD* pronephric duct, *TB* tail bud)

Chimaeric combinations	Hosts (*n*)	Survivors	Structures identified in sections						
			NT	NC[a]	ME	OP	PD	GL	TB
Crist→Taen	124	42 (34%)	18	13	12	2	7	3	2
Crist→Alp	72	12 (16.5%)	5	4	3	0	0	2	0
Others	63	19 (30%)	3	2	2	0	0	2	0
Total	259	73 (28%)	26	19	17	2	7	7	2

[a] Some embryos showing a secondary neural plate were not sectioned and therefore the presence of a notochord could not be established, or their donor material was not shaped like a notochord. The remaining structures listed at the head of the table were chimeric or entirely of host origin

her baby son's food when she spilled some of the liquid and within seconds stood ablaze. Refusing help from her mother-in-law, she rushed out to her husband, who quickly extinguished the flames, but too late. She died the next morning, on 5 September 1924. Her tombstone still stands in Gotha (Fig. 8).

Hilde Mangold's organizer experiments

The original laboratory notes and slides of Hilde Mangold[6] permit a much more detailed record of her work than any previous publication (see Fässler 1991, 1995). During the spring months of 1921 and 1922 she performed 259 heteroplastic transplantations for her thesis. This figure equals the number of hosts, while that of donors is considerably lower because 30 donors provided more than one graft each (see, for example, Fig. 10 below). These experiments yielded "secondary" axial structures at the belly or flank of 28 host embryos (Table 1). In 26 of these, the secondary structures were partly – and to different degrees – composed of host cells. The secondary structures ranged from bona fide neural plates to the complete pattern of axial organ rudiments, lacking only the head parts anterior to the otic vesicles.

The overall yield of inductions thus was close to 10%. However, the various newt species were differently suited for providing host embryos. The nearly unpigmented eggs of *Triton* (now *Triturus*) *cristatus* were rarely used as hosts because of their "touchiness" when handled, perhaps aggravated by the lethal factor discovered much later in that species (Wallace 1987). This is unfortunate for the photographic record, because pigmented graft cells (especially from *T. alpestris*) can be recognized at first glance in an unpigmented host (Fig. 9a,b) while unpigmented cells are less easily spotted in a pigmented host. This holds specifically for the most viable combination (Table 1), *T. taeniatus* hosts (which are weakly pigmented) with grafts from *T. cristatus* (Fig. 9c,d). Among the 124 chimaeras of this constitution, 18 yielded chimaeric secondary structures consisting at least partly of host cells. That is, every sixth transplant effected some axial induction. Naturally the percentage is even higher when one excludes the embryos that had died before neurulation: the 18 secondary axes induced in 42 surviving hosts amount to 43% – a yield that compares favourably with famous cytoplasmic injection series recently performed in *Drosophila*! The surviving *T. alpestris* hosts gave comparable yields, but survival as such was much lower in this species (Table 1). Thirteen homoplastic transplantations using *T. taeniatus* (control experiments – not included in Table 1) gave the best survival (100%), but astonishingly only a single secondary axis.

Of the 26 chimaeric secondary embryos available when Hilde wrote her thesis in the autumn of 1922, only six are described and discussed in the organizer paper of 1924. That must have been Spemann's decision, because common experience (shared by both Ph.D advisers and editors) suggests that Hilde would have loved to see some more cases in print. But between them these six cases showed all "the basic facts" (as stated in the Introduction) of the organizer effect revealed by her experiments.

We cannot detail the unpublished experiments here (see Fässler 1991 and 1995 for the complete record), but attention might be drawn to some cases that differed in approach. A very interesting one was not mentioned in the paper. In this experiment (Fig. 10), three grafts were taken from a single dorsal lip. The outcome was extremely fortunate when measured against the average yield, because all three grafts induced secondary structures in their respective hosts, and each of these obligingly survived! The laboratory notes do not record the motives that led to this particular experiment. On the face of it, the aim would have been to see whether or not grafts taken far off the midline would still be effective; this question was followed

[6] These materials are kept in the Embryological Collection of the Hubrecht Laboratory, Netherlands Academy of Sciences, Utrecht. They were listed incognito under Otto Mangold – even Hilde's slide boxes carried his name, unknowingly attached by the staff – until identified by K.S.

Fig. 9a, b Chimaeric secondary axial structures demonstrating the organizer effect. **a** Experiment 1923 Um 168. The *T. alpestris* graft had been rotated by 180° and consequently did not invaginate but sank below the surface. *Top* The photomicrograph (325x) shows *T. alpestris* pigment granules in the notochord and an adjacent part of paraxial mesoderm while the remainder of the mesoderm and the neural tube consist of unpigmented *T. cristatus* cells. *Bottom* Pencil rendering of the whole section, drawn by Hilde Mangold in 1923 (with identification lines added by Otto Mangold for the 1929 publication). The pigmented *T. alpestris* cells are distributed as in the photograph. **b** Experiment 1922 Um 132. The unpigmented *T. cristatus* graft was internalized mainly by sinking into the weakly pigmented *T. taeniatus* host. *Top* In this famous figure taken from Spemann and H. Mangold (1924) the notochord and the adjacent portion of the somite are unpigmented, as are the floor plate cells. *Bottom* In the photomicrograph (380x), this is best seen when these structures are compared to the remainder of the left somite and to the lateral walls of the neural tube, where the pigment has concentrated in the apical region bordering the lumen. (Photographs taken in 1991 by K.S. using the Leitz Ortholux/Orthomat equipment)

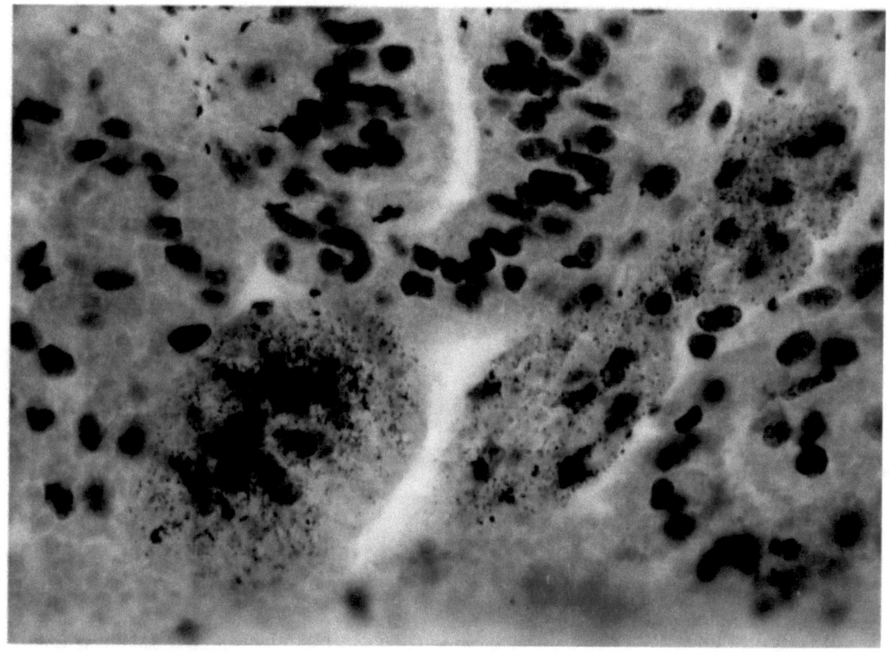

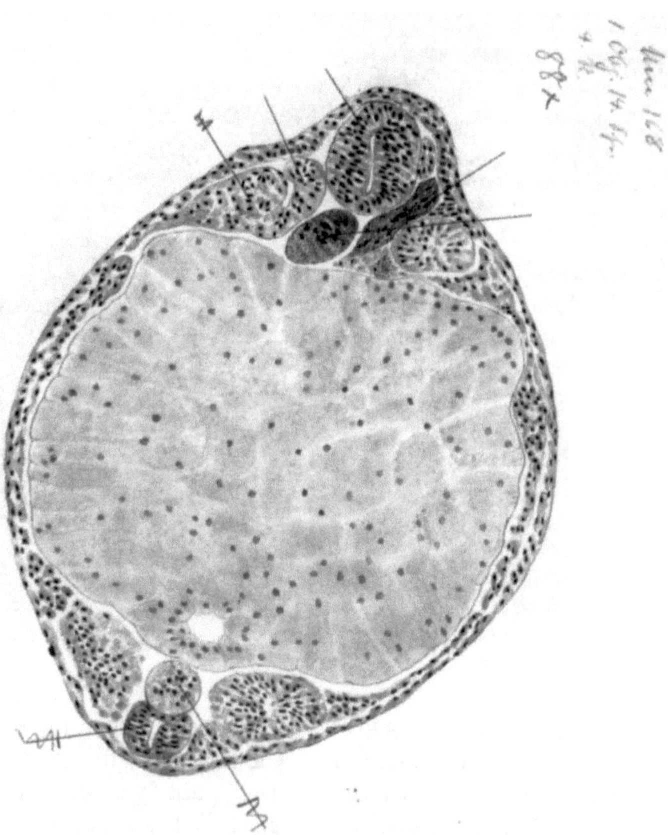

a

up in 1923 (see below). However, given the preoccupation of early experimenters – notably Hans Driesch, but also Hans Spemann (e.g. 1936, p. 223) – with "wholeness", the true question might have been whether wholeness can be split up in amphibians by this method, and to what degree (see note added in proof). Another question was whether the transplant would carry along its original polarity. This was discussed in the 1924 paper, but the task of answering it was referred to future experimentation. In fact, however, Hilde had in 1922 tried to answer that question, by using triangular transplants; but apparently the outcome was too ambiguous for publication.

The results jointly obtained by Hilde and Otto Mangold in the spawning season of 1923 (Fässler 1991) were also not incorporated in the 1924 paper. Besides Spemann's selectivity and the problem of co-authorship, there may have been a formal reason for excluding work in which Otto Mangold was involved. According to the Ph.D. statutes, the candidate's thesis had to be published in print (and this, of course, without the addition of subsequent data); the 1924 paper was evidently to serve this function.[7] Moreover, the

[7] A reprint was actually kept for this purpose at the university's central library, in a somewhat worn cardboard cover that stated as much in handwriting. When told that this was an interesting document related to a Nobel prize, the first thing the librarian did was to have the cover replaced by a new one. The typed version submitted by Hilde in 1923 does not seem to have survived.

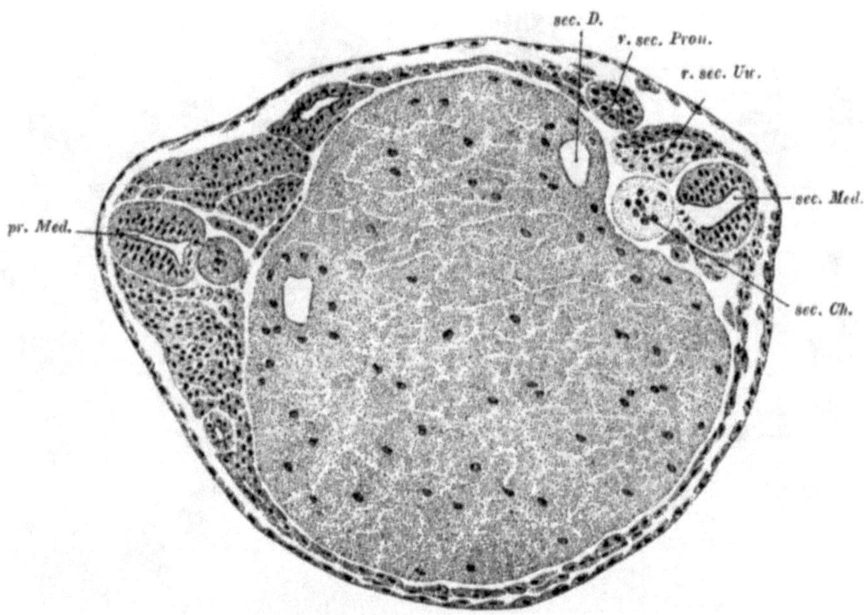

Abb. 24. Um 132 b. Querschnitt durch die Mitte des Keims (vgl. Abb. 21). Auf Abbildung links die primären, rechts die sekundären Achsenorgane. *r. sec. Pron.* rechter sekundärer Vornierengang. Implantat (hell) als Chorda und im rechten sekundären Urwirbel. 100 ×.

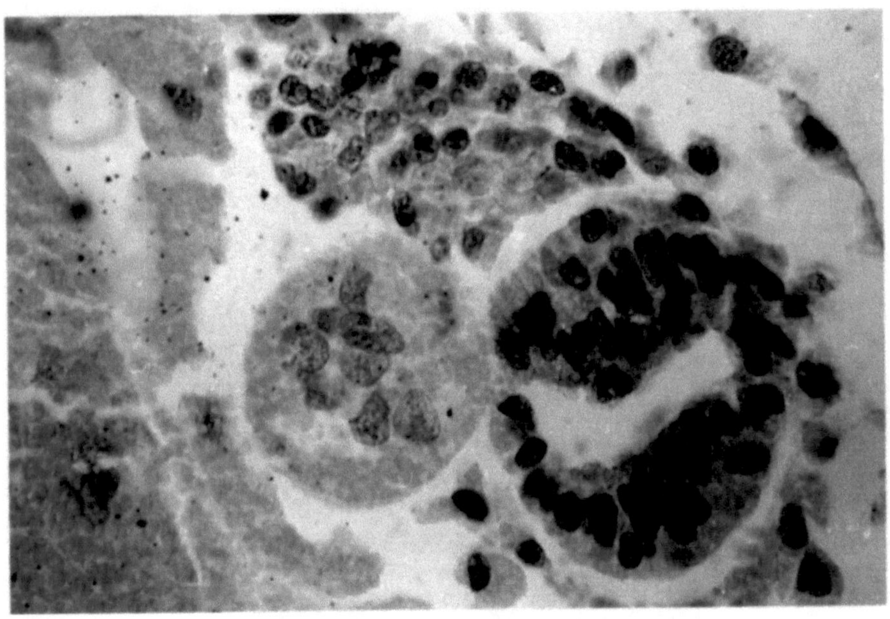

b

1923 experiments aimed at answering some more specific questions, mainly concerning the spatial extent of the organizing capacity in the embryo and the influence of transplant size. They comprised 217 hosts, of which a mere 17 developed secondary structures. Several factors may have contributed to this low yield (7% of the total), including a large proportion of sensitive hosts (mainly *T. alpestris*) as well as transplants taken from near the limits of the organizer region.

Taken together, the positive results known by the time when the 1924 paper was submitted can be classified as follows. Among the 43 hosts carrying an induced secondary neural plate or tube, a notochord was documented in 35 cases, flanked in 29 of these by at least one row of somites or pre-somitic mesoderm. These structures were often somewhat deformed and less numerous or shorter than their normal counterparts. Moreover, as criticized by Jacobson (1982) – and in contrast to the specimens depicted by W.H. Lewis (see below) – the induced structures generally had not reached the stage where they could be identified by histotype. However, each of them looked so different from anything else in embryos of the respective stage, and so much like its (assumed) counterpart in the host's own axis, that the experienced eye could classify them with a high degree of reliability – as subsequently proven by countless transplantation experiments in other laboratories. Still lacking, on the other hand, was proof that the head structures anterior to the otic vesicles can be induced as well by dorsal lip material; this proof was soon to be provided by Spemann and Geinitz (1927. p. 168).

Critical questions answered

Critical questions about the experiments and interpretations of Spemann and Hilde Mangold (1924) were occasionally raised on at least two accounts: (1) low yields (e.g. Horder and Weindling 1986) implying an insufficient number of cases (cocktail party gossip only), and (2) doubts about the origins of the secondary neural plates and about the inferred neural character of their cells (Jacobson 1982). While the latter doubts were soon dispelled (Gimlich and Cooke 1983; Jacobson 1984), the former should be silenced by the figures given in Table 1 – note that they certainly surpass the yield of some famous germ line transplantations in *Drosophila*!

Another topic occasionally vented at cocktail parties is whether Hilde Mangold really was the first to demonstrate the organizer effect. An early suggestion to the contrary came from Gavin R. de Beer (later Sir Gavin), who in 1927 wrote: "It must not be forgotten that while the detailed analysis of these phenomena is the work of Spemann and his school, the credit of making the first organizer graft belongs to Lewis (1907b). He grafted the organizer from a gastrula of *Rana palus-*

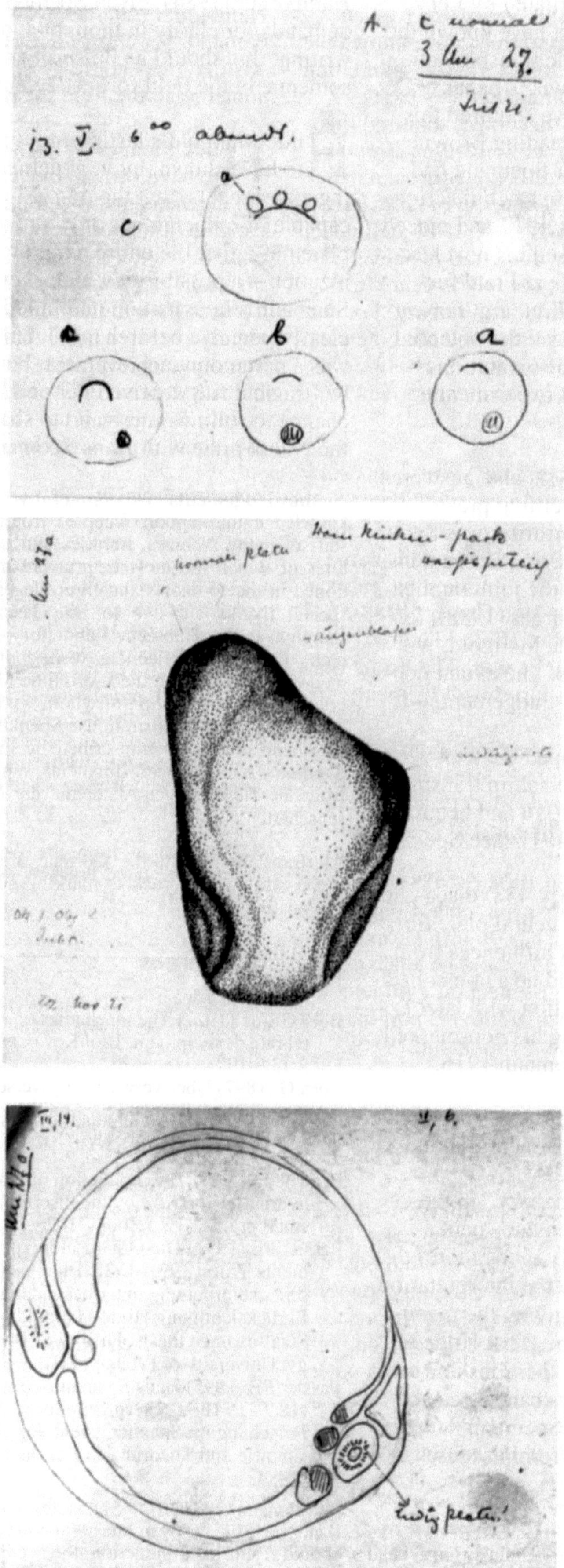

Fig. 10 Laboratory notes and drawing from experiment UM 27 of 13 May 1921. *Top*: Three transplants taken from the dorsal lip of a single *T. cristatus* donor were each transplanted into a *T. taeniatus* host. *Middle*: Ink drawing from Hilde Mangold, showing a lateral view of the host of graft "a", which has two somewhat bent axes (the induced one is on the *right*). *Bottom*: Laboratory sketch of a section through the same host, with the secondary axis at *bottom right*; the parts derived from the graft are *hatched*. Reproduction of the laboratory notes courtesy of the Hubrecht Laboratory, Utrecht; the ink drawing is kept at the author's institute

tris into the otic region of another embryo in the tailbud stage, and obtained an induction of nerve tube, notochord and somites."

Sven Hörstadius (1928, p. 101) went one better by succinctly stating some months later: "The first proof of an organizing region was provided in 1907 by Lewis." Yet Lewis (1907) came no closer to this than to suggest that in a single case a secondary notochord might have caused the adjacent epidermis to thicken (no hint here at *neural* induction!). As for the secondary neural tubes and muscles he had observed in about ten hosts, he would have been unable to provide proof that they were induced because he had no means of distinguishing between the respective contributions of graft and host. What was even more important: Lewis had no incentive for suspecting induction of the secondary structures he observed,[8] because at that time everyone – Morgan and Spemann included – held the opinion

[8] It is possible that, indeed, no induction was involved, because (at least in newts) a host at the tailbud stage will no longer respond to a lip transplant by forming a well-developed secondary axis; thus, in Lewis' experiments the secondary structures might all have arisen from the graft. In any case, Lewis' aim was "to determine the extent of independent self-differentiation possessed by this structure [i.e. the lip] when small pieces were isolated and transferred into strange environments". In the end he opined that "the dorsal and lateral lips of the blastopore are already determined as regards their subsequent differentiation to chorda, muscle, and nervous tissue... " (1907, p. 140).

that in normal development the neural plate, somites and notochord all originate by self-differentiation from the lip material and its surroundings (see Sander 1991, 1996). So it was easy for Spemann to quip at his detractors that "one may well make a discovery without intending to, but not, it seems to me, without realizing it" (Spemann 1931, p. 512). In fact, during his long and distinguished career Lewis never claimed to have discovered the organizer effect – not even when the temptation to claim this might have been greatest, namely when soon after the Second World War he hosted a postdoctoral student from Otto Mangold's laboratory (Renate Rollhäuser-ter Horst, personal communication).

Matters are somewhat different as regards Ethel Browne (later Ethel B. Harvey). She is said to have – on a single occasion – claimed credit for having discovered the organizer (Lenhoff 1991). Whether she actually did claim this or not, she had certainly realized that something peculiar was going on in *Hydra* (cf. Fig. 5) when a small implant from the hypostome caused an almost complete secondary body to be produced from the gastric wall (Browne 1909). Hence Lenhoff raised the question whether Browne rather than Hilde Mangold deserved credit for discovering the organizer phenomenon. Browne, however, discussed her findings under the heading of regeneration, using the long-established term induction (see Oppenheimer 1970) but once; she never spoke of organization or the like. Moreover then, as now, most people would have considered the induction of a hydra less intriguing than the riddles of nascent vertebrate organization. All this readily explains why Browne's work (in contrast to Hilde's) failed to ring in a new epoch in developmental biology. The use of grafts that differed in pigmentation from the host – which permitted Browne to prove induction – also caused no sensation because this approach had long been used in amphibian research (Born 1897; Harrison 1903). Therefore (and contrary to a possibility ventilated by Lenhoff) Spemann must have known the benefits of chimaeric transplantation before reading Browne's paper. Hilde, too, must have acquired this knowledge before reading Browne (if she ever did), for Spemann had made spectacular use of chimaeric newt embryos since 1917 and indeed should have discussed his novel experiments in the talk at Frankfurt that lured Hilde to Freiburg. For certain he had recognized the potential of chimaeric lip transplantations before delegating that experiment to Hilde (see below).

Finally, our analysis also answers a question concerning precedence rather than priority: was Spemann justified in having his name precede Hilde's on the joint publication? Viktor Hamburger (1988), fellow student of Hilde Mangold, thought that he was. The extant documents support this judgement:

1. Spemann developed the techniques required: dorsal lip transplantation in newts in 1916 and heteroplastic grafting in 1917 (see Spemann 1919).
2. He had in 1904 (p. 458) discussed the possibility that neural plate formation depends on influences from the archenteron, and he again considered this in connection with later experiments including his non-marked lip transplants (Spemann 1916, 1918).
3. He had shown the way by which to test the concept of neural induction (Spemann 1921).
4. According to his diary,[9] in the spring of 1923 Spemann spent at least 50 h (probably much more) discussing with Hilde the interpretation of her results and giving the manuscript its final shape. That Hilde fell somewhat short of these tasks when preparing her thesis can be read between the lines of Spemann's report (Fig. 6) and in the reminiscenses of Hamburger (1984); given Spemann's demands for clarity in thought and writing, this should be no surprise in someone in the field for just 2 years!

Hilde Mangold's lasting merit is to have shown, by very demanding experiments, that the dorsal lip is capable of inducing not only a neural plate but rather the entire axial organization – a possibility which even Spemann's imagination had failed to clearly perceive beforehand.[10] This was a paramount achievement, but her tragic death deprived her of the chance to follow it up – and to share the Nobel prize with Hans Spemann.

Acknowledgements We gratefully acknowledge the support received from the staff of many archives, libraries and institutes of which the most helpful are mentioned in the footnotes and figure legends; special thanks are due to Dr. Gesineke Bangma at the Hubrecht Laboratory, Utrecht. Personal recollections were contributed especially by the niece of Hilde Mangold, Rose Lakschewitz, to whom we also owe some of the illustrations. Spemann's shorthand diary was transcribed by Franz Kremp, Freiburg, with financial support from the Heidelberger Akademie der Wissenschaften.

All translation from the German are by K.S., with square brackets marking added explanations.

References

Beer GR de (1927) The mechanics of vertebrate development. Biol Rev (Camb) 2:137–197
Born G (1897) Über Verwachsungsversuche mit Amphibienlarven. Arch Entwicklungsmech Org [Roux' Arch] 4:349–465, 517–623
Browne E (1909) The production of new hydranths in *Hydra* by the insertion of small grafts. J Exp Zool 7:1–23
Fässler PE (1991) Die Organisatorexperimente Triton 1921–1923. Eine historische Aufarbeitung unter besonderer Berücksichtigung Hilde Mangolds. Staatsexamen thesis, Faculty of Biology, University of Freiburg i.Br.
Fässler PE (1995) Hans Spemann (1869–1941) – experimentelle Forschung im Spannungsfeld von Empirie und Theorie. Ein Beitrag zur

[9] Spemann at that time was *Rektor* (the elected academic and administrative head of the university) and, because of the responsibilities inherent in this task, kept daily notes by the hour.

[10] Spemann 1921 (p. 551) states clearly that, even if the secondary neural plate should result from induction, the accompanying notochord *and* somites were to derive from the transplanted piece of the blastoporal lip.

Geschichte der Entwicklungsphysiologie zu Beginn des 20. Jahrhunderts. Ph.D. thesis, University of Freiburg i.Br.

Gimlich RL, Cooke J (1983) Cell lineage and induction of second nervous systems in amphibian development. Nature 306:471–473

Hamburger V (1984) Hilde Mangold, Co-discoverer of the organizer. J Hist Biol 17: 1–11

Hamburger V (1988) The heritage of experimental embryology. Hans Spemann and the organizer. Oxford University Press, New York

Harrison RG (1903) Experimentelle Untersuchungen über die Entwicklung der Sinnesorgane der Seitenlinie bei den Amphibien. Arch Mikroskop Anat Entwicklungsgesch 63:35–149

Horder T, Weindling PJ (1986) Hans Spemann and the organiser. In: Horder T, Witkowsky JA, Wylie CC (eds) A history of embryology. Cambridge University Press, Cambridge

Horder T, Witkowsky JA, Wylie CC (eds) A history of embryology. Cambridge University Press, Cambridge

Hörstadius S (1928) Über die Determination des Keimes bei Echinodermen. Albert Bonnier, Stockholm

Jacobson M (1982) Origins of the nervous system in amphibians. In: Spitzer N (ed) Neuronal development. Plenum Press, New York, pp 45–99

Jacobson M (1984) Cell lineage analysis of neural induction: origins of cells forming the induced nervous system. Dev Biol 102: 122–129

Lenhoff HS (1991) Ethel Browne, Hans Spemann and the discovery of the organizer phenomenon. Biol Bull 181:72–80

Lewis WH (1907) Transplantation of the lips of the blastopore in *Rana palustris*. Am J Anat 7:137–143

Mangold H (1929) Organisatortransplantationen in verschiedenen Kombinationen bei Urodelen. Ein Fragment, mitgeteilt von Otto Mangold. Wilhelm Roux's Arch Dev Biol 117:697–711

Mangold O (1932) Das Determinationsproblem. III. Das Wirbeltierauge in der Entwicklung und Regeneration. Erg Biol 7:193–403

Mangold O (1953) Hans Spemann, ein Meister der Entwicklungsphysiologie. Sein Leben und sein Werk. Wissenschaftliche Verlagsgesellschaft, Stuttgart

Nieuwkoop P (1969) The formation of the mesoderm in urodelean amphibians. I. Induction by the endoderm. Wilhelm Roux Arch Entwicklungsmech Org 162:341–373

Oppenheimer JM (1970) Some diverse backgrounds for Curt Herbst's ideas about embryonic induction. Bull Hist Med 44:241–250

Sander K (1991) When seeing is believing. Wilhelm Roux's misconceived fate map. Roux's Arch Dev Biol 198:177–179

Sander K (1996) Vertebrate gastrulation: unity in teratogenesis? Axial clefts in a historical perspective. Neth J Zool 46:134–146

Spemann H (1904) Ueber experimentell erzeugte Doppelbildungen mit cyclopischem Defect. Zool Jahrb Suppl VII (Weismann-Festschrift). Fischer, Jena, pp 429–470

Spemann H (1916) Über Transplantationen an Amphibienembryonen im Gastrulastadium. Sitzungsber Ges. Naturforsch Freunde Berl 1916:306–320

Spemann (1918) Über die Determination der ersten Organanlagen des Embryo. I–IV. Arch Entwicklungsmech Org [Roux' Arch] 43:448-555

Spemann (1919) Experimentelle Forschungen zum Determinations- und Individualitätsproblem. Naturwissenschaften 7:581–591

Spemann H (1921) Die Erzeugung tierischer Chimären durch heteroplastische embryonale Transplantation zwischen *Triton cristatus* und *taeniatus*. Arch Entwicklungsmech Org [Roux' Arch] 48:533–570

Spemann H (1931) Über den Anteil von Implantat und Wirtskeim an der Orientierung und Beschaffenheit der induzierten Embryonalanlage. Wilhelm Roux's Arch Dev Biol 123:389–517

Spemann H (1936) Experimentelle Beiträge zu einer Theorie der Entwicklung. Springer, Berlin

Spemann H, Geinitz B (1927) Über Weckung organisatorischer Fähigkeiten durch Verpflanzung in organisatorische Umgebung. Wilhelm Roux's Arch Dev Biol 109:129–175

Spemann H, Mangold H (1924) Über die Induktion von Embryonalanlagen durch Implantation artfremder Organisatoren. Arch Mikrosk Anat Entwicklungsmech [Roux' Arch, see footnote 1] 100:599–638

Wallace H (1987) Abortive development in crested newt *Triturus cristatus*. Development 100: 65–72

Willier BH, Oppenheimer JM (1964) Foundations of experimental embryology. Prentice-Hall, Englewood Cliffs, N.J.

Note added in proof

A relevant printer's error might be of interest from the human point of view. It concerns Driesch's "harmonious-equipotential system", conceived for safeguarding wholeness ultimately on a vitalistic basis. On p. 634 of the organizer paper, the authors refer to the analytical *Unzugänglichkeit* (inaccessibility) of this concept – as rendered correctly in the translation of Willier and Oppenheimer (1964, p. 180) – but the text reads *Unzulänglichkeit* (insufficiency). That this was overlooked at proofreading has a Freudian touch to it, considering the deep-rooted differences between Speman and Driesch.

Roux's Archives
a century in
Developmental
Biology

Centennial Essay

An American in Paris and the origins of the stereomicroscope

Klaus Sander

Abstract. A century ago, Horatio S. Greenough, an American living in Paris, persuaded the firm of Carl Zeiss to construct the first low-power stereomicroscope. Fitted by Zeiss with Porro prisms for image erection, this instrument became the ancestor of all stereoscopic dissection microscopes ("binoculars") now in use. On the basis of original documents and earlier publications, the present Centennial Essay traces both the history of the Greenough stereomicroscope and the scantily documented life of its inventor, including his probable descent from a family of famous New England sculptors. The aim is to provide some historical background for a truly timeless instrument that was fundamental to developmental biology, beginning almost from the days of its proclamation by Wilhelm Roux.

Key words: Horatio S. Greenough – Stereomicroscope – Dissecting microscope – Binocular – Carl Zeiss

Introduction

Besides the centenary of its first journal, known since 1925 as Roux's Archives, developmental biology in 1994 ought to commemorate yet another innovation – one of which memories seem to have been largely lost. The candidate is the Greenough stereomicroscope (Fig. 1), named after Horatio S. Greenough (pronounced Greeno), an American who lived in Paris. In several respects this instrument is the direct ancestor of every stereomicroscope (dissecting microscope) in use today. All in all, its impact on nascent developmental biology was probably no less important than that of Wilhelm Roux's *Archiv für Entwickelungsmechanik*.

Fixing a date for the Greenough centennial is admittedly difficult, but 1994 is as good a choice as any, because just a century ago this particular stereomicroscope was halfway through its gestation period. Conception, as far as written records show, occurred in 1892 with the fusion of American inventiveness and German experience and craftsmanship, whereas birth followed towards the end of 1896; a pre-series model was tested by Braus and Drüner (1895). If the selection of 1994 for the Greenough centennial should need further justification, one might muster the first commercially successful prism field glasses (called *Feldstecher*), which came on the market exactly one hundred years ago. The two instruments have crucial features in common, and both were introduced by the firm of Carl Zeiss at Jena in Thüringen.

The "theory" of his instrument was outlined by Greenough some years later (manuscript draft of 1901) "in a semi-popular form": the stereomicroscope was to act as if the observer's eyes were sliding down along its

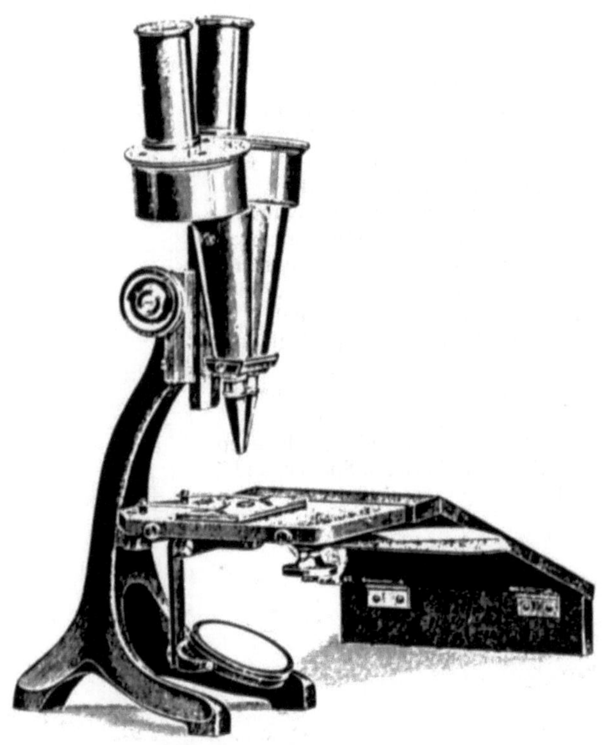

Fig. 1. "Greenough's stereomicroscope" of Carl Zeiss, earliest model (except for the pre-series). (From Czapski 1897)

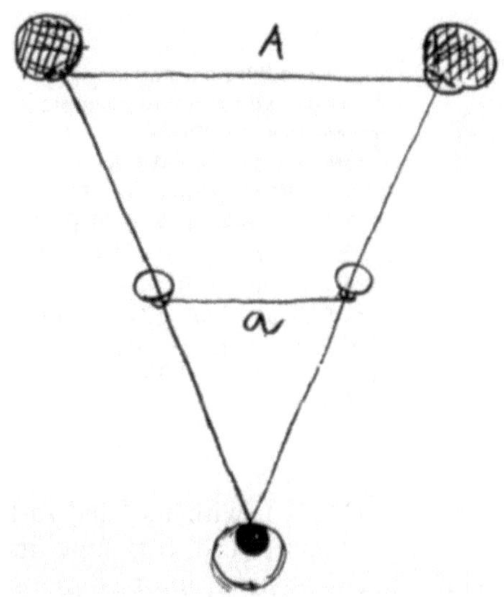

Fig. 2. The principle of microscopical stereoscopy; sketch by Horatio S. Greenough, showing the eyes of the observer and of the "dwarf"; A and a represent the respective interocular distances. (From a draft of 1901, Thüringisches Hauptstaatsarchiv, Weimar)

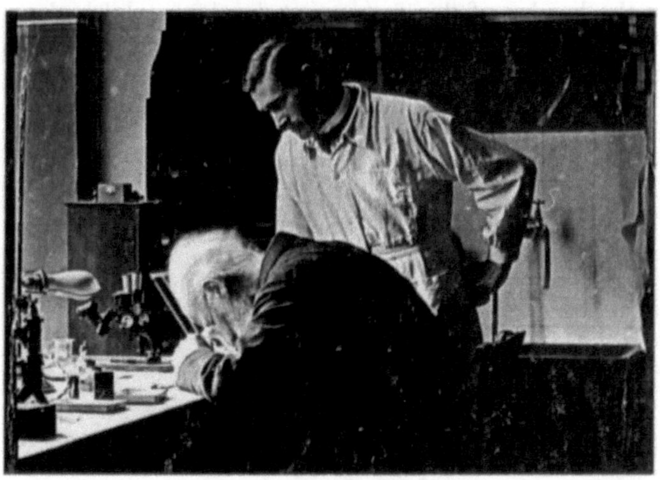

Fig. 3. August Weismann, apparently bent over a dissection lens. In front of his crown, a Zeiss-Greenough can be seen detached from its stand and stage. (Photograph probably around 1910, copy in the author's institute)

convergent optical axes, shrinking proportionally as they progressed, until they would be gazing from quite close – like the eyes of a dwarf – at the object in focus (Fig. 2).

Before the stereomicroscope

How did experimentally inclined embryologists get along before the advent of the stereomicroscope? Well, most of them exploited techniques accessible without low-power stereoscopic observation, such as oxygen deprivation, piercing, shaking or centrifuging of eggs, or exploitation of reproducible natural mishaps like polyspermy (see Oppenheimer 1968; Sander 1990 and the Essays in recent issues of this journal). But some others, foremost among them Abraham Trembley and Henry Baker in the eighteenth century, performed admirable feats of microsurgery with the most primitive means (Ford 1985). Just

Fig. 4. Zeiss dissection stand carrying a Brücke-type compound microscope, erecting the image by means of a concave eye lens. (From Zeiss 1870)

consider Trembley isolating a Hydra tentacle for microscopical observation (Lenhoff and Lenhoff 1986): peering through a magnifying glass of focal length between 11 and 14 mm (magnification should be about 20–25 times), he persuaded a Hydra to attach the tip of its arm to a fine brush, by which he could draw it gently out of the water, thereby extending the arm since the animal's foot kept clinging to its support. After having picked up with the other hand a narrow strip of glass, "all that remains to be done is ... to manoeuvre the slide under this arm until they touch. Then, by making a small motion with the hand that holds the slide, I break the arm off on both edges of the slide." This sounds simple – but just try doing it yourself!

Somewhat more comfortable for dissection work were the "single lens microscopes" of the eighteenth and nineteenth centuries, which sported an object stage and a lens – often of the doublet type – focused by pinion and rack. According to the expert opinion of Brian Ford (1985), they owe their invention directly to Henry Baker's need for a better view on his "polypes". All these "aquatic" and "botanical" microscopes, however, had the disadvantage of short working distances, the specimen being located next to the lens and – more awkwardly – to the experimenter's nose: witness the aged August Weismann bent over the desk of a young collaborator (Fig. 3). But this photograph from the first decade or so of our century also signals a change for the better: for the young man himself apparently had at his disposal the Zeiss Greenough microscope seen to the left of Weismann's head (Weismann may not have used it because of his almost life-long eye troubles). The working distance problem can be illustrated by yet another document. When Carl Zeiss, in 1870, introduced a small *Präparir-Mikroskop* of the Brücke type – intended to replace the single lens on dissecting stands

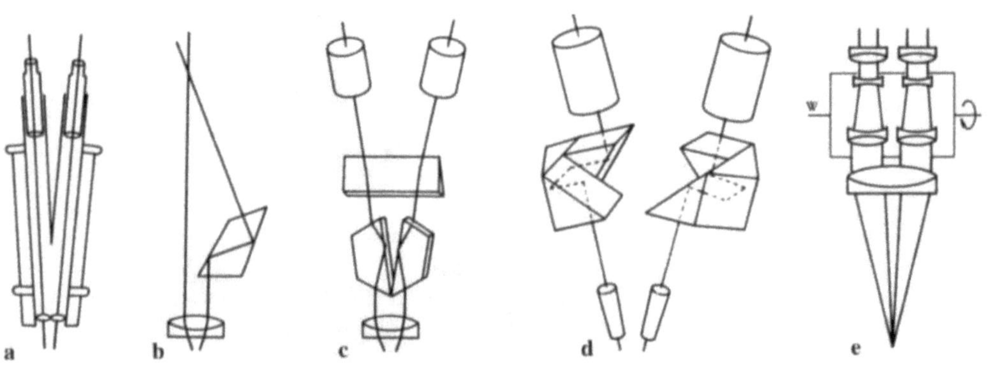

Fig. 5a–e. Optical pathways in various designs of stereoscopic microscopes. a Cherubin d'Orleans, 1671. b Wenham 1860. c J.W. Stephenson 1870. d H.S. Greenough as modified by Zeiss (cf. Fig. 7) (1894; shown is the modified prism set of about 1920). e Design of Carl Zeiss, Jena (1940s). (From Greiner and Sander 1987)

(Fig. 4) – he announced proudly that owing to the stretch of brass tube replacing the short mount of the single lens "the eye is assigned the very comfortable elevation of 70–80 mm [i.e. about 3 in.] above the stage".

Early stereomicroscopes

Although obviously inconvenient in use, these dissecting microscopes provided the upright image required for manipulating tiny objects. Yet that image was not stereoscopic, owing to its monocular perception. Binocular compound microscopes were constructed – at least in theory – quite early (see Wise 1949; Greiner and Sander 1987; Beck 1988), beginning with the honourable Cherubin d'Orleans in the seventeenth century (Fig. 5a). But his instrument must have produced a pseudoscopic rather than an orthoscopic image, inverting spatial depth so that a ball-shaped object in focus would be perceived as a dish.

Stereoscopy was not even considered in microscope construction until after Charles H. Wheatstone had made known (in 1838) the principles of binocular depth perception. In subsequent decades, several inventors achieved orthoscopic stereoscopy by means of a prism (or rather, in most cases, a set of prisms) that split the beam behind the single front lens of a compound microscope fitted with two diverging tubes. The commercially most successful of these instruments was the large Wenham model of 1860, which was fitted with a unilateral prism (Fig. 5b) at the bases of a skewed pair of shiny brass tubes; for many decades this was the British microscopist's pride and favourite conversation piece, but, to quote the famous American pathologist Charles Minot (1877), "the abhorrence of histologists".

Before the advent of "the Greenough", the nearest approximation to modern stereomicroscopes relied on twin prisms placed just above the objective, followed by "erecting prisms" somewhere further up the optical pathway (Fig. 5c). This image-erecting and orthoscopic arrangement was conceived and realized independently by the American chemist J.L. Riddell in 1853 and the British insurance mathematician J.W. Stephenson in 1870. The former – and the "amazing effects" he had claimed for his instrument – was soon forgotten, while Stephenson is mainly remembered (at least outside the British Isles) for advising Ernst Abbe to construct immersion objectives to be used with highly refractive oils rather than water. The Stephenson instrument (Fig. 6), with its hand rests and inclined tubes, was convenient to use. It is suprising that it did not meet with much enthusiasm among developmental biologists (the present writer at least is not aware of any developmental studies mentioning it). The reasons may hail from the halving of aperture caused by the prisms and from the still fairly short free distance between front lens and focal plane. It is just as likely, however, that in the heyday of this instrument developmental biology was not yet ripe for sophisticated experimentation on embryos.

All this changed with the advent of the Greenough instrument, which we shall consider now.

Fig. 6. Stephenson binocular microscope of 1870. The manufacturer's claim in 1910 that this is "undoubtedly the most perfect binocular microscope" was probably not accepted at that time by the majority of developmental biologists. It may have been tenable, however, because its higher magnifications (which were not possible with the Greenough principle) were helpful "for the selection and arrangement of diatoms" in ornamental patterns, to be admired under high power – a favourite pastime of amateur microscopists and commercial slide producers. (From the 1910 catalogue of James Swift & Son, London)

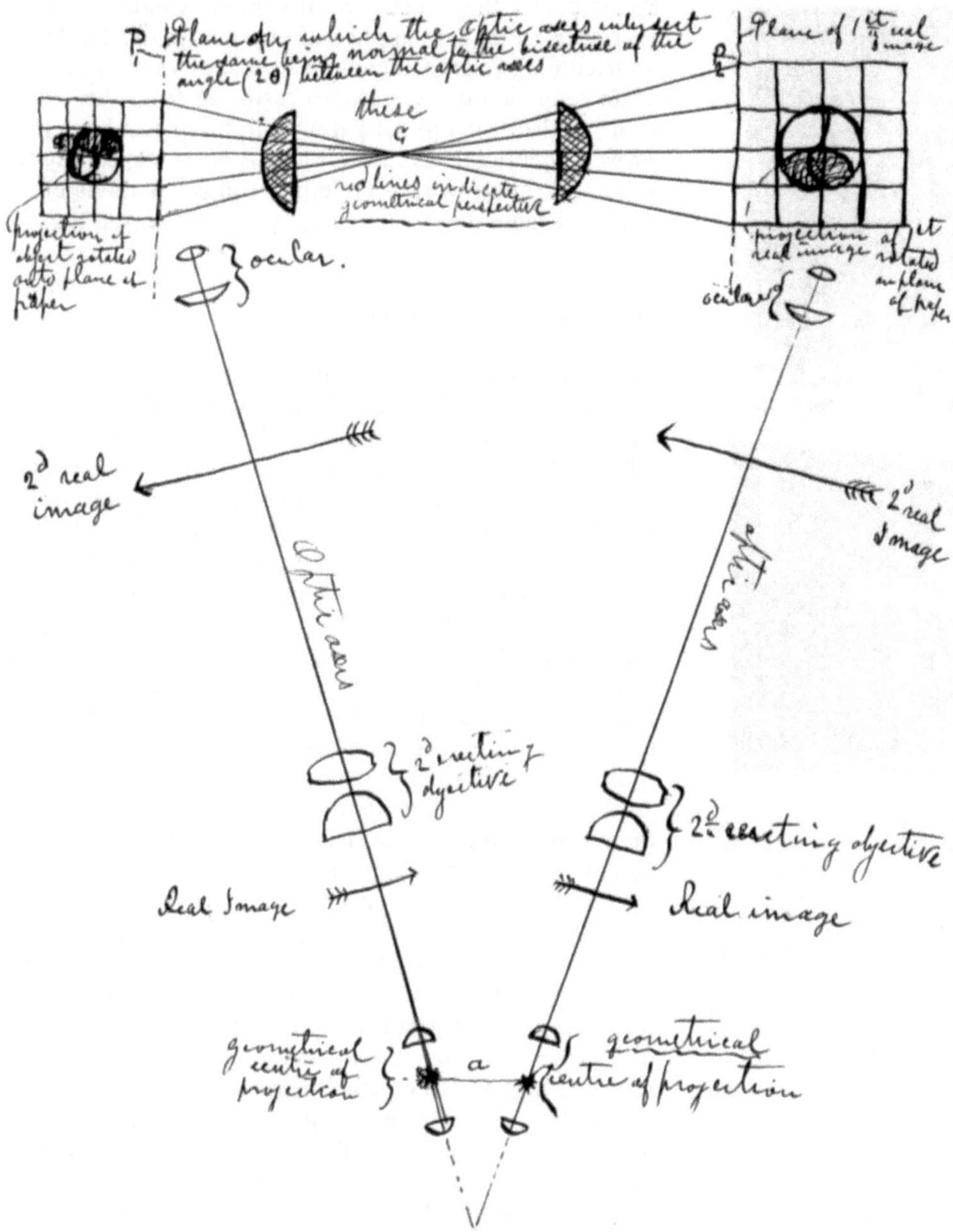

Fig. 7. Sketch by Horatio S. Greenough of his concept for an orthomorphic stereomicroscope, from a letter of 20 September 1897; note the "2nd erecting objective" within each tube, for which Zeiss substituted the Porro prism sets. This type of objective was used in terrestrial telescopes before; Pieter Harting (1859) had embodied it in his design for a stereomicroscope, but in order to increase magnification rather than to erect the image. (Original in the Thüringisches Hauptstaatsarchiv, Weimar)

The origins of the Greenough stereomicroscope

It all started with a letter from the Laboratoire de Zoologie Maritime at Concarneau in Brittany, dated 6 September 1892 and written by one Horatio S. Greenough to Ernst Abbe, the head and brains of the firm of Carl Zeiss until shortly before. In this letter, Greenough explained his plans for a double microscope with converging tubes that was intended to convey an orthoscopic and orthomorphic image. The sketch accompanying this letter seems to be lost, but the Zeiss files contain another drawing in Greenough's hand, which was sent to Abbe's successor Siegfried Czapski in September 1897 (Fig. 7). This shows clearly that Greenough – contrary to a statement made by Wise (1949) – planned to erect the image of his microscope by means of a second pair of objectives that would invert the images produced by the first. He suggested – and throughout more than a dozen years repeatedly demanded – that Zeiss should build the instrument this way. Incidentally, an almost identical concept had been proposed by Pieter Harting in 1859. Yet Greenough may well have arrived at the idea independently, such duplication of ideas being not uncommon in optics: witness the Riddell/Stephenson coincidence and the use of Porro prisms in field glasses (see below). In any case, it was Greenough and not Harting who communicated the principle to Zeiss and thereby made microscope history.

Greenough was apparently obsessed with the idea of orthomorphy, that is, the notion that the microscopical object should be seen "dilated" to the same degree in all three dimensions of space. Orthomorphy was also behind his additional postulate that the object should be kept (by the fairly long tubes) at reading distance from the eyes. Greenough claims that these conditions were met by his

Fig. 8. Ross G. Harrison at his bench, using a Zeiss "Stand Xa" Greenough with the straight upright supplied between 1912 and 1923. On the window sill a dissection stand can be discerned, with its single lens turned to the left, perhaps for checking cultivated embryos. Copied from a lantern slide dated 26 January 1918 – one in a series apparently given by Harrison to his friend Hans Spemann. (Original in the author's institute)

design. The experts at Zeiss acknowledged this but refuted Greenough's scheme because of various mechanical and optical problems associated with it. Instead they recommended – and in the end incorporated – the very Porro prisms (see Fig. 5d) that were just being tested in the *Feldstecher*. Abbe had arrived at this prism set independently and only later found, and acknowledged, that it had been conceived for prism telescopes about 1850 by I. Porro but not implemented at the time owing to technical shortcomings. With these erecting prisms, the perceived depth of the object was not strictly proportional to its horizontal "dilation", but they had the crucial advantage that the distance between the eyepieces could be changed by simply swivelling the prism boxes around the tubes' optical axes. This made adaptation to the user's interocular distance much simpler (and optically "cleaner") than in all previous binocular microscopes – for which reason this mode is the only one to have survived from the nineteenth century stereomicroscopes to the present day.

For orthomorphy to be achieved with Greenough's concept as modified by Zeiss, a certain correction was needed, namely placement of the entrance pupils of the objectives closer to the object. The conical stops seen to protrude downward from the short, drum-shaped objective mounts in Fig. 1 served this purpose. But their use reduced optical resolution, so that even Greenough himself recommended removing the stops for dissection work; indeed they were soon discarded by most users and therefore no longer supplied after a few years. The other shortcoming (from Greenough's point of view), the moderate reduction of tube length resulting from the "doubling up" of the optical pathway in the prism box, soon proved beneficial (as in the *Feldstecher*!), especially for microdissection work – including transplantations etc. – which became the instrument's true domain (see Fig. 8).

Documentation of success

From its inception, the "binocular microscope after Greenough", as modified by Zeiss, proved highly successful. This is well documented by written evidence, in addition to the ever rising sales (and speedy copying of the instrument's design by other firms). Braus and Drüner (1895), working in the anatomical institute at Jena (next door, so to speak, to the Zeiss workshops), had been hankering for a test model while the firm of Zeiss was still hesitant to start production – probably giving priority to the market demand for the new *Feldstecher*. When they finally got their test model, Braus and Drüner used it for tracing fine nerves in vertebrate dissections but remarked that "it also promises great benefits for other investigations (for instance the observation of egg cleavage)." Accordingly, on introducing the instrument to the public in their 1898 catalogue, the firm of Zeiss announced: "This new binocular microscope is to facilitate the observation of small three-dimensional objects such as seeds, crystals, eggs, larvae etc. by means of a realistic, three-dimensional rendering thereof in an erect image, and therefore should provide a welcome aid foremost to zoologists, botanists and embryologists." This prediction apparently came true soon, for in that same year, 1898, H. Harting reported that the Greenough instrument of Carl Zeiss had met with "great applause in all circles" – so much so that the firm immediately decided to develop four additional objective pairs considerably extending the range of magnifications (originally 25× to 40×, now 12× to 72×); the stronger ones stood out below the tubes like the "orthomorphic stops" in Fig. 1 but of course they carried lenses at their tips, not just apertures.

Among the early users writing about the instrument was Hans Spemann. It is probably not by chance that he started his egg ligations and similar experiments in 1897 (Fig. 9), the year when "the Greenough" had just become available commercially; rather, Spemann must have learned early on about this instrument from his close friend Braus (see Fig. 9) – and he certainly knew what problems to attack with it. When in 1907 he detailed his by then famous embryological micro-tools and their use, he wrote: "All these subtle handlings would hardly be possible without Greenough's fine binocular which, as is known, was introduced for dissection purposes by Braus and Drüner who adapted it very practically to this purpose." A more specific statement was made by Warren Harmon Lewis at Johns Hopkins University (1904), who by transplantation experiments within and between different frog species had documented positively – several years before Spemann – that the eye cup is capable of

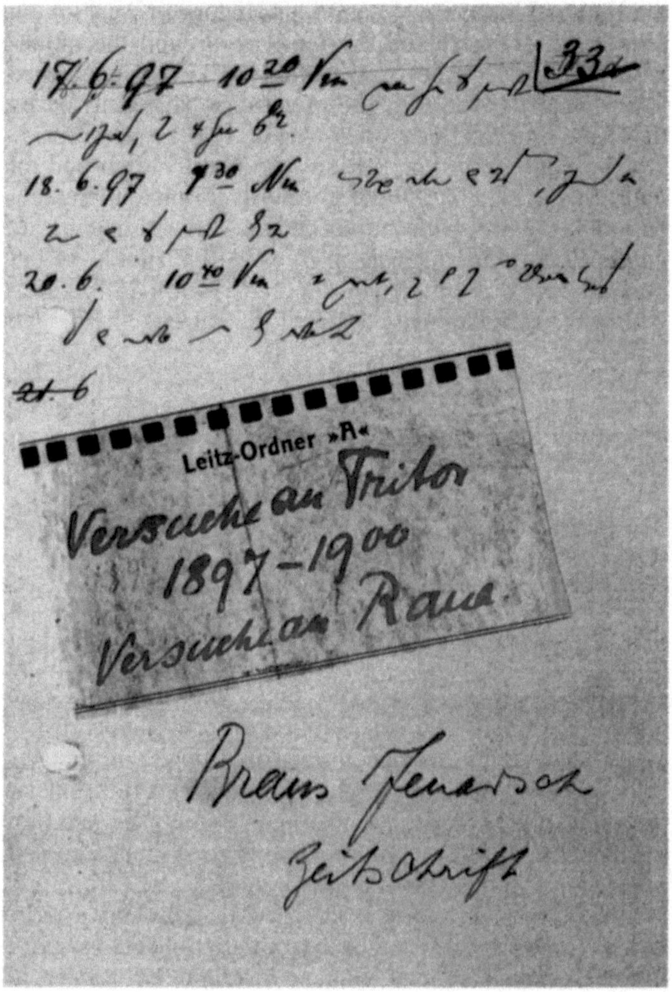

Fig. 9. Notes on a ligaturing experiment carried out on *Triton* (now *Triturus*) *taeniatus* by Hans Spemann in 1897; superimposed is the label of the respective file holder. The sheet carries a reference, in Spemann's hand, to the Braus & Drüner (1895) report on Greenough's stereomicroscope. (Original in the author's institute, reproduction 2/3 size)

inducing ectopic lens formation: "By use of the binocular microscope" – which, considering the year of publication, should have been a Greenough – "one can make minute dissections of the living amphibian embryo and can remove various organs, transplant them or alter the normal relations, and so alter the influences they exert on each other. We may thus determine certain correlations necessary to normal development" – a central quest of embryologists since the programmatic writings of Wilhelm Roux (see Sander 1991).

Horatio S. Greenough – his character and life

Very little is on record about Greenough the man, almost everything known being based on his correspondence with the Zeiss staff (see Gause 1965). In one letter, he claims to have had Professor Benjamin M. Pierce as his teacher at Harvard, but little is known of his achievements as a scientist: after decades of hunting for publications from Greenough's pen, I have come up with just two short notes (Greenough 1891, 1892). One of these is dealing with embryological techniques and mentions some haphazard results obtained with these. The other compared gastrulation in amphibians and *Peripatus* (Onychophora), concluding that in both groups the nerve cord is homologous. For a few years around 1892, Greenough was listed as a member of the Société Zoologique de France; in its register he is said to hold membership in the Société de mathématiques et de physique de Boston. He mentions having met C.S. Minot and C.O. Whitman in the U.S.A. In a (lost) letter to which Zeiss responded in 1906, he apparently claimed to have been made an "Officier d'Academie" – a title that, I am told, was conferred by the government, not a scientific body. Greenough is known to have worked at Concarneau where Y. LeGal (personal communication) located his name in the visitors' register of 1889, and from where he wrote to Abbe in 1892. According to his own testimony, it was there that he received from Laurent Chabry the impetus for a "capillary rotator" (Sander and Fischer 1992). Like his "prism rotator", this was produced for some years by Zeiss and later on by Spencer; both implements were devised in order to permit microscopic observation of embryos from different angles, the former by turning them enclosed in a capillary, the latter by reflecting their hidden lower or lateral surfaces. In 1897/98 he mentions working facilities at the Ecole Normale Supérieure, provided apparently by Professor Frederic Houssay.

That Greenough is largely lost to history probably has its main roots in his personality. His permanent change of Paris addresses (four alone in the Zeiss correspondence, plus one at Neuilly sur Seine where he apparently settled for good before 1906), his trips to fashionable shooting and fishing resorts with ever changing friends, his short membership in the Société Zoologique, and last but not least the extreme vacillations of his attitude towards the Zeiss experts (even reflected by marked variations in his handwriting) all speak of a restless and unstable personality torn between attempts at friendly social interaction and recurrent phases of hostile captivity in his pet ideas – namely the orthomorphic stereomicroscope and a "new axiom in mathematical philosophy". He claims that this axiom – "Invariance of the proximate efficient cause is the necessary and also the sufficient condition to the operation of the principle of least action" – had found its manifestation in his optical concept (letter of 17 December 1896). The permanent refusal by the firm of Zeiss to produce his true "Orthomorphoscope" (Fig. 7) beyond a workshop model caused Greenough to virtually forbid the company to use his name with their instrument, and this on at least two separate occasions (1897 and 1902). These outbursts were followed within days, in one instance by telegraph, by apologies and reconciliation; the Zeiss people always reacted as patiently as with a child, even trying to assist him in matters of health – personal ties that Greenough seems to have cherished even during otherwise antagonistic periods.

The forces at work in Greenough are perhaps best grasped from a letter he wrote to Lord Avebury of the Royal Society, dated 3 November 1902 and sent in copy to

Zeiss: "The Carl Zeiss people are aware that the microscope listed by them under my name meets with my sanction simply because I wish to act in fairness to them; but I have long since informed them that I consider the one they make to be an inefficient makeshift, and that the listing of the same under my name might injure my reputation as a man of science." His emotional sufferings are revealed by a letter to Czapski and von Rohr, dated 31 December 1906 (quoted here with his spelling and emphases): "And I request that as soon as you can you should make public statement that *this form* of [microscope] stand is Agreed to by me because preferred by *you* to the *Integral Execution of my own design* – by so doing you will relieve me of acute moral distress of many years standing."

Part of Greenough's personality structure may have hailed from his family background, to which this writer gained a (hopefully correct) lead by a rather unusual route. On being asked about Greenough, the keeper of a famous microscope collection admitted to not knowing the name, let alone anything about the man, but his secretary spoke up: "Isn't that the man of the marble statue of George Washington at the Smithsonian?" This 12-ton monument – commissioned by the U.S. Congress in 1832, as the first such offer to an American sculptor – is indeed associated with that very name. Greenough the famous sculptor lived in Italy while executing this work and for some years thereafter. According to Gause (1963), the microscopist's father was said to have been a rather well-known sculptor and writer – which would fit Horatio S. Greenough Sr., who indeed is also remembered for his writings as an ardent protagonist of functionalism (Gardner 1945; Metzger 1970; see also entry "Greenough" in the Encyclopaedia Britannica). But our quest was not to be that easy, for the sculptor/writer's brother Richard S. (=Saltonstall) Greenough (1819–1904) was a sculptor, too, and also lived in Europe (Italy, Paris) for a long time; the only way we have of differentiating the two is that Richard was not known as a writer (nor was a third Greenough brother, again a sculptor but of minor fame). Interestingly, Horatio Jr., in two letters to Zeiss, had recourse to the sculptor's approach when advocating orthomorphic imagery. The best bet, then, is that the "Inconnu of the Stereomicroscope" was the son of one of these artist brothers – which could simultaneously account for his interest in orthomorphy, his passionate character, and his preference (and that of his sister, who was married to a Frenchman) for living in Europe. The connection to a wealthy New England family background, which would follow from this line of argument, might moreover have provided him with the means of supporting his apparently very independent life style.

A clue to Greenough's mother (and hence to his father) might hopefully come from his own second initial "S.", which (unless added because of the sculptor's fame) should be that of his mother's family name. In this case, Horatio Sr. would be excluded from paternity, because in a letter (see Greenough 1887) he mentions getting married to a Miss Gore.

As for the dates of birth and death of Horatio S. Jr., nothing is known for sure. However, the administration of the National Museum of American History in Washington (which was constructed, by the way, around the monumental marble statue of Washington) kindly unearthed from their files the following Obituary Note, carried by the New York Times of 5 April 1916: "Horatio S. Greenough, a native of Boston who had lived abroad for many years, died on Sunday at Neuilly, France, in his seventy-first year." Unless two people with the same rare name lived simultaneously at Neuilly, this should be our man. His birth could then be dated to 1845 or 1846 – well within the reproductive phases of the various Greenough sculptors.

Subsequent development of the stereomicroscope

Zeiss produced their Greenough stereomicroscope for almost four decades without much change, apart from replacing the original set of prisms by another configuration (actually that shown in Fig. 5d, also conceived by Porro). The design was copied, as was usual at that time (and sometimes is still today), by several optical firms in England, France, Germany, and the USA (see Greiner 1986). While some optical improvements, especially wide-field eyepieces and high-power objectives (12×), were first introduced by German firms, the Americans (Spencer, Bausch and Lomb) took the lead in mechanical improvement. Besides two widely adopted devices for speedy switching of objectives (drum and revolver), they soon encased individual objective pairs in a common shell, a much needed and obvious way of protecting them from mechanical disalignment – a chronic weakness of the Zeiss instrument which the writer and his fellow students cured (back in the austere postwar period) by slinging a rubber band around the pair of "noses", thus steadying one objective by tying it to the other.

The first basic departure from the Zeiss-Greenough concept (Fig. 5e) became apparent after World War II, when the two branches of Zeiss, soon to be separated by the Iron Curtain, introduced their "Citoplast" and "Opton" stereomicroscopes – obviously of common ancestry hailing from Jena. These instruments (many of which must still be in service) had roof-angle erecting prisms, a single large front lens corrected for infinity, and a corresponding pair of tube lenses. This permitted maintenance of the very large working distance through the full range of magnifications. The novel means of changing magnifications was a safely encapsuled drum (axis marked W in Fig. 5e) carrying two pairs of Galilei-type telescopes and a pair of void bores. Depending on the drum's position, the telescopes would either increase or decrease the magnification provided by the front lens. Easily recognized by its characteristic five-step magnification range, this construction is still on the market today – as is (from several smaller firms) the venerable Zeiss-type "Greenough" fitted with Porro prisms. Thus, ironically or – from the human point of view – rather tragically, the name of Horatio S. Greenough is kept alive by a design that he despised from its very beginnings.

Acknowledgements. My heartfelt thanks are due to the many people who at one time or another generously supported my quest for Greenough the man and his stereomicroscope, foremost among them my former student Kurt Greiner (Maulburg), who wrote the first truly comprehensive history of stereomicroscopy, on which I have drawn heavily. Much additional information concerning Greenough and his instrument came from documents kindly made accessible by the Zentralarchiv of JENOPTIK Jena GmbH and the Thüringisches Hauptstaatsarchiv Weimar, from the Billings Microscope Collection at the Walter Reed Medical Center near Washington, and from Mr. Ebenezer Gay of the Harvard Historical Instruments Collection, all of whom I wish to thank at this point. I am also indebted to the Wissenschaftliche Gesellschaft and the Verband der Freunde der Universität, both in Freiburg i.Br., who at times provided financial support for this project. The author's institute maintains a collection of historical stereomicroscopes, for which offers of obsolete instruments (or parts thereof, like objectives, stands etc) and printed documents would be welcome at any time.

References

Beck R (1988) 75 Jahre Leitz-Binokularmikroskope. Leitz Mitt Wiss u. Technik 9: 133–145

Braus H, Drüner L (1895) Ueber ein neues Präpariermikroskop und über eine Methode, größere Tiere in toto histologisch zu konservieren. Jena Z Naturwiss 29 (NF 22): 435–442

Czapski S (1897) Das stereoskopische Mikroskop nach Greenough. Z Wiss Mikrosk 14: 289–303

Ford B (1985) Single lens – the story of the simple microscope. Harper & Row, New York

Gardner AT (1945) Yankee stonecutters – the first American school of sculpture. Metropolitan Museum of Art/Columbia University Press, New York

Gause H (1963) Siegfried Czapski 1861–1907. In: Freund H, Berg A (eds) Geschichte der Mikroskopie, vol III., Umschau-Verlag, Frankfurt/M, pp 109–117

Grenough FB (ed)(1887) Letters of Horatio Greenough to his Brother, Henry Greenough. Ticknor, Boston (Reprinted 1970 in the Library of American Art, Da Capo Press, New York)

Greenough HS (1891) Observations sur les larves d'oursin. Bull Soc Zool Fr 16: 239

Greenough HS (1892) Sur les homologies des premiers stades suivant la segmentation chez les batraciens. Bull Soc Zool Fr 17: 57–59

Greiner K (1986) Stereomikroskope und Binokulartuben: Ihre Entwicklung und ihre Anwendung in der Biologie bis 1950. Typescript Staatsexamen-Thesis, Zoologisches Institut, Universität Freiburg i Br

Greiner K, Sander K (1987) Das Stereomikroskop – Ursprünge und geschichtliche Entwicklung. Biol in unserer Zeit 17: 161–168

Harting P (1859) Das Mikroskop. Vieweg, Braunschweig

Harting H (1898) Über einige optische Vervollkomnungen an dem Zeiss-Greenough'schen stereoskopischen Mikroskop. Z Wiss Mikrosk 15: 299–303

Lenhoff HM, Lenhoff SG (1986) Hydra and the birth of experimental biology – 1744. Boxwood Press, Pacific Grove

Lewis WH (1904) Experimental studies on the development of the eye in Amphibia. Am J Anat 3: 505–506

Metzger CR (1954) Emerson and Greenough – transcendental pioneers of an American esthetic. University of California Press, Berkeley (Reprinted 1970 by Greenwood Press, Westport/Conn)

Minot CS (1877) The study of zoology in Germany. Amer Naturalist 11: 330–336 and 392–406

Oppenheimer J (1968) Some historical relationships between teratology and experimental embryology. Bull Hist Med 44: 241–150

Sander K (1990) From germ plasm theory to synergetic pattern formation: a century of developmental modeling in retrospect (in German). Verh Dtsch Zool Ges 83: 133–177

Sander K (1991) Wilhelm Roux and his programme for developmental biology. Roux's Arch Dev Biol 200: 4–7

Sander K, Fischer JL (1992) How to dart ascidian blastomeres: the embryological micro-tools of Laurent Chabry (1855–1893). Roux's Arch Dev Biol 201: 191–193

Spemann H (1907) Über eine neue Methode der embryonalen Transplantation. Verh Dtsch Zool Ges 196–202

Wise FC (1949) The development of the binocular microscope. J Queckett Microscopical Club, ser 4, vol II 6 (Reprinted as Monograph of the same Club, Williams and Norgate, London 1950)

Zeiss C (1870) Ein neues Präparir-Mikroskop. Arch Mikrosk Anat 6: 234–236

Roux's Archives
a century in
Developmental
Biology

Centennial Essay

Is ventral in insects dorsal in vertebrates?

A history of embryological arguments favouring axis inversion in chordate ancestors

K. Nübler-Jung, D. Arendt

The idea that the ventral side of insects and other arthropods corresponds to the dorsal side of vertebrates has been put forward from time to time since early in the nineteenth century. It was Etienne Geoffroy St.-Hilaire who, in 1822, originally proposed that the arthropod body plan is like the vertebrate body plan turned upside down. His claim was virtually torn apart by Georges Cuvier but re-emerged, with increasing recourse to embryological arguments, about every 50 years thereafter, only to be rejected time and again. However, as we have argued elsewhere, recent molecular and embryological data may be in its favour. Here we take the opportunity, provided by this journal's centennial volume, to review the observations and arguments of earlier embryologists who tried to compare insect and vertebrate development by inverting the dorsoventral axis in one of the groups, whether just formally or – after Darwin's *Origin of species* – in the assumed course of evolution. The pre-evolutionary writers mainly discussed are M.H.Rathke, R.A.Kölliker and G.Zaddach. From the period after the *Origin*, we review the work and thoughts of F.Leydig, C.Semper, A.Dohrn and A.Naef.

Key words: Chordate evolution – Annelid hypothesis – Animal body plans – *Unité de plan* – La querelle des analogues (1830).

Introduction

The evolution of chordates from lower animal groups remains to be clarifield to this very day. More than 150 years ago, Geoffroy St.-Hilaire (1822) proposed that arthropods and vertebrates represent a common body plan, with the exception that arthropods chose to walk with their dorsal side oriented towards the ground. Since then, it has been suggested that the chordates derive, respectively, from crustaceans via arachnoids (Gaskell 1898), from arachnids (Patten 1890) or from annelid-like organisms via dorsoventral inversion (Leydig 1864; Semper 1874; Dohrn 1875; Naef 1926a; see Fig. 1). Other worm-like organisms such as nemertines (Balfour 1878, as quoted by Gaskell 1898), Turbellaria (Goette 1895), the Hemichordata (Bateson 1886) or, more recently, worms with a metameric coelom (Gutmann 1967, 1976) have been proposed as chordate ancestors, as have organisms like the auricularia larva (Garstang 1894), free-living oligomeric organisms (Archicoelomata, Remane 1967), the sessile Tentaculata (Romer and Parsons 1977) or the bilaterally symmetrical carpoids (Jefferies 1991).

Interestingly, recent molecular data show that even rather distantly related animal groups utilize similar genetic mechanisms to lay down their morphologically divergent body plans: annelids, arthropods, nematods and vertebrates all display a similar longitudinal pattern of gene expression, specifically of the *Hox* cluster genes (the "zootype" of Slack et al. 1993). On this basis, we have proposed that chordates and articulates (i.e. annelids and arthropods) may share a common ancestor with a ventrally situated nerve cord. Comparing ventral and dorsal expression patterns of various homologous genes

Diagram of a notoneuralian (vertebrate) with brain (*x*), dorsal nerve cord (*n*), and with the blood streaming anteriorly in the ventral vessel (*arrows*). The anterior gut does not pierce the brain. The mouth forms by fusion of the stomodeal invagination (*s*) with the anterior gut. (*nc* notochord, which does not exist in gastroneuralian animals). Modified textbook diagram, see for example Romer and Parsons (1977)

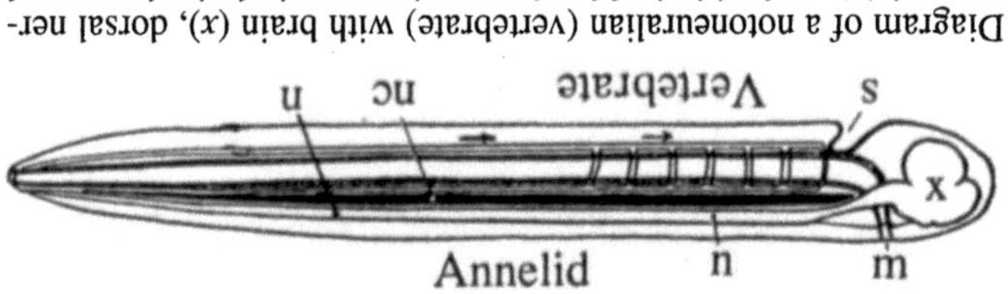

Fig. 1. Diagram of a gastroneuralian animal (e.g. annelid) with brain (*x*), ventral nerve cord (*n*) and ventral mouth (*m*). Note that the anterior gut passes through the brain. *Arrows* at the dorsal vessel indicate anteriorly directed blood flow. To compare with notoneuralian organization, turn page upside down.

Offprint requests: K. Nübler-Jung

in insects and vertebrates, we have suggested previously that the nerve cords of chordates and articulates both evolved from the nerve cord of their last common ancestor, and that the ventral side of articulates corresponds to the dorsal side of chordates (Arendt and Nübler-Jung 1994).

While pondering on this idea, we realized that it had been put forward several times since the days of Geoffroy St.-Hilaire, only to be rejected time and again. Recent authorities on phylogenesis also consider the idea of evolutionary axis reversal unattractive (e.g. Romer and Parsons 1977; Starck 1978; Kaestner 1993). We believe, however, that the apparently conserved colinear expression patterns of *Hox* cluster genes in articulates and chordates, together with some very striking ventral versus dorsal expression patterns of specific genes, now provide the incentive as well as additional tools for re-evaluating this old theory. In this Essay, we shall review the non-molecular approaches mainly for the historical record (which explains our listing some certainly abstruse details). However, recording them may also prove useful for future debates on the topic, because the old authors defined, and may have solved in part, several problems that arise from the idea of a dorsoventral inversion during chordate evolution.

For this overview we have drawn on the work of various other authors. For the ideas of Geoffroy St.-Hilaire and Cuvier we took recourse to Appel (1987), Hassenstein (1958), Jahn et al. (1985) and Tort (1983). Some of the information on Rathke comes from Zaddach (1854) and from Kölliker's latin thesis (1842), which was translated into German by Wieland and Sander (see Wieland 1993). Leydig's and Semper's results are reported by Kühn (1950), who also described Anton Dohrn's scientific work and the ensuing controversy with Gegenbaur and Haeckel. The life of Dohrn was masterly rendered by Theodor Heuss – later to be the founding president of the Federal Republic of Germany – whose book is now available in English (see Heuss 1940). An English translation of Dohrn's fundamental work is also about to appear (see Dohrn 1875). Where we were unable to refer to an original publication, we have listed our secondary sources in the References. Omissions from and additions to quoted phrases are set in square brackets.

Geoffroy St.-Hilaire versus Cuvier: Are all animal groups based on a common body plan?

More than 150 years ago, Etienne Geoffroy St.-Hilaire (1772–1844), comparative anatomist and embryologist in Paris, proposed that all body plans in the animal kingdom, despite their large morphological differences, are variations of a single common plan (*unité de plan*, Geoffroy St.-Hilaire 1822). For example, he suggested that the gill arches of fishes correspond to the larynx of higher tetrapods (accepted today) and that the operculum of bony fish corresponds to the ossicle(s) of the tetrapod middle ear (not accepted today). However, Geoffroy St.-Hilaire not only compared various structures within the group of vertebrates, but ventured to compare the body plans of the different animal phyla. For example, he proposed that the vertebrate body plan corresponds to that of a crayfish (*Astacus*) turned upside down. In this orientation the main organ systems are essentially arranged from bottom to top as in the vertebrate body. In addition, Geoffroy proposed that the vertebrae correspond to the body segments of an arthropod, and the split spinal arches of vertebrates to an arthropod's legs. Arthropods would thus have encased their soft parts *within* their "vertebrae", while vertebrates developed their body mainly *around* their vertebrae (Geoffroy St.-Hilaire 1822; for further details see Appel 1987; Hall 1992).

Georges Cuvier (1769–1832), comparative anatomist, palaeontologist and outstanding colleague of Geoffroy St.-Hilaire (as well as Lamarck) in Paris, vehemently opposed Geoffroy's ideas of a unified body plan. Cuvier was convinced that the animal kingdom comprised four quite separate groups, namely the *Vertebrata*, *Articulata*, *Mollusca* and the *Radiata* (mainly the coelenterates and echinoderms). He maintained that the specific body plans of these groups were quite discrete and therefore could *not* be compared (Cuvier 1817).

These very divergent views eventually clashed in 1830. When two of Geoffroy's students attempted to deduce a common body plan for cephalopods and vertebrates, there arose *la querelle des analogues*, the famous public debate between Geoffroy and Cuvier. The latter, by enumerating more differences than similarities between duck and squid (Jahn et. al. 1985), decided the argument in his favour. Johann Wolfgang von Goethe (1749–1832), the famous poet, despite his quite advanced age took a lively interest in the debate and tried to point out the analytical versus synthetical approaches of the opponents; but with his inclination towards unifying concepts, Goethe tended to side with Geoffroy St.-Hilaire (Goethe 1830a; see also Hassenstein 1958; Tort 1983; Appel 1987).

Rathke (1829) and Kölliker (1842): Embryogenesis reveals that the ventral side of arthropods corresponds to the dorsal side of vertebrates

Geoffroy St.-Hilaire's idea of a unified body plan for the large animal phyla remained a challenge throughout the nineteenth century. Almost every description of arthropod development in that period closed with a discussion of how far the embryos of arthropods and vertebrates (at that time represented mainly by the chick) could be compared to one another.

For example, Martin Heinrich Rathke (1793–1860, comparative anatomist in Königsberg) investigated the anatomy of *Amphioxus* – which up to then had been considered a snail's larva – as well as the development and anatomy of crustaceans, molluscs, worms and vertebrates. After a thorough examination of the development of the crayfish *Astacus*, Rathke concluded that the early development of articulates and the chick is very similar. Both form an oblong "primitive part" (pars primitiva, germband or primitive streak) later on flanked by leg buds. The primitive part later expands laterally around the yolk and thereby forms the body cavity. Since in the chick the primitive part represents the dorsal body side

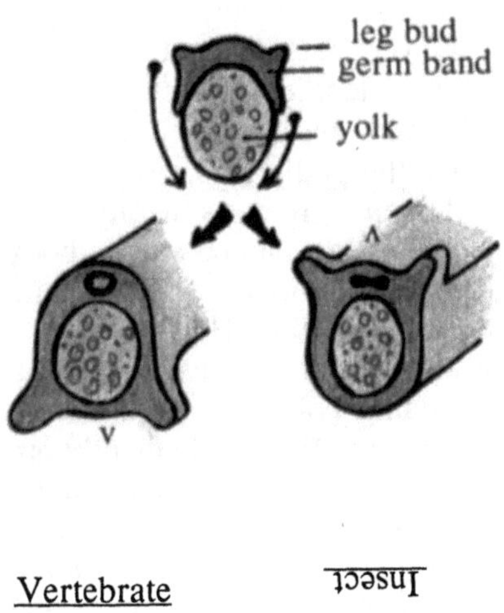

Fig. 2. Diagram representing Rathke's comparison (1829) of early development in a vertebrate (chick, *left*) and in an insect (*right*). In both groups the flanks of the primitive streak/germ band overgrow the yolk (*arrows*). During later development, in vertebrates (*bottom left*) the legs protrude from the side opposite to the nervous system, while in insects (*bottom right*) they remain on the body side where the primitive streak had formed. The dorsal side of a vertebrate would thus correspond to the ventral side of an insect. v = ventral

and its flanks grow ventrally while in arthropods it represents the ventral side with the flanks growing dorsally, Rathke concluded that the ventral side of artropods corresponds to the dorsal side of vertebrates (Fig. 2). He stressed, however, that the later development of articulates and vertebrates is very dissimilar, beginning with the fact that in articulates the legs remain close to the nerve cord while in the chick they come to lie on the opposite side. Since articulates never form a neural tube, Rathke concluded that the nerve cords of articulates and vertebrates, in spite of their similar position in the midline of the former germ band, cannot be considered equivalent structures (Rathke 1829, as rendered by Kölliker 1842).

It is noteworthy that even before Darwin published his *Origin of species* (1859), the similarities between insects and vertebrates were often interpreted to mean that vertebrate embryos tend to pass through a more primitive arthropod-like stage before developing towards the higher organization of vertebrates. This view was expressed, for instance, by Rudolf Albert Kölliker (1817–1905), famous anatomist and cytologist from Switzerland, who taught mostly in Würzburg. In his medical dissertation, he described the course of embryogenesis in several insects, notably chironomids and beetles (Kölliker 1842). He closed by discussing similarities between arthropods and vertebrates, stating that "developmental history so far has given the most valuable testimony for evaluating organs in their correspondence or differences, and with its support I shall carry that which I have begun to a fortunate end."

Kölliker repeatedly expressed his conviction that the (functionally) ventral side of articulates is, in reality, dorsal because – according to his criteria – it entirely corresponds to the dorsal side of vertebrates. Since arthropod legs thus develop on the *dorsal* side, it is quite natural that arthropods turn this side downwards. He supports his view by quoting divergent postures seen in some other animals, for instance the flat fish resting on one side, and several crustaceans which turn towards the ground their abdomen (*Daphnia*, *Polyphemus*), their dorsal side (*Branchipus*) or their lateral side (amphipods).

Specifically, Kölliker claimed that arthropod and vertebrate nervous systems are initially similar in shape and position, consisting of two rows of ganglia which later fuse. The arthropod nerve cord then remains solid while the vertebrate nerve cord goes on to develop into a tube, a sequence that is no surprise – "since we all know that an organ reflects its shape in lower animals once again during its early development in higher animals". Kölliker even believed that the ventral longitudinal muscle strands of arthropods represent "the vestiges" of notochord and vertebrae which "are found in the same position in vertebrates and, like the former, originate from the primitive part where it borders the inner layer."

Kölliker then noted that in articulates the foregut departing from the mouth pierces the *Primitivstreifen* (and therefore is later flanked by the circumpharyngeal connectives), while in vertebrates it ultimately opens on the side opposite the nerve cord. He found the situation in insects easy to explain – "since the mouth parts develop there, the mouth for complete harmony must develop there, too". He concluded that in the germband stage of arthropods and vertebrates "almost everything is similar, if we designate the primitive streak of articulates and vertebrates to be the same, and posit that the belly of one corresponds to the back of the other." Yet, he continued, "there remains a certainly difficult matter when it comes to comparing and interpreting the limbs of both."

In order to solve the problem of appendage positioning, Kölliker took a rather unorthodox view. Under the premise that ventral in arthropods corresponds to dorsal in vertebrates, he described how in arthropods the appendages develop in the position occupied in vertebrates by the *Rückenplatten* (upright dorsal plates which would later fuse to form a tube; for possible pitfalls of the definition, see legend to Fig. 4). These longitudinal elevations were to be segmentally interrupted in arthropods, yielding the individual appendage buds. "Therefore", he concluded, "arthropods must lack that hollow tube which [according to von Baer] in vertebrates forms by upfolding and fusion of the dorsal plates". But he pointed out that one can find "already in arthropods – if only in some of them – indications of the vertebrate condition, in that the right and left dorsal plates of the *Simulia* larva have anteriorly fused into a single medial appendage, and therefore here the dorsal plate behaves just as in vertebrates" (Fig. 3). Conversely, appendages comparable to those of arthropods usually do not develop in vertebrates, except in fish, whose dorsal fins Kölliker interpreted to be an outgrowth of the medially fused rims of their dorsal plates (Fig. 4).

Because arthropod legs correspond to the dorsal plates of vertebrates, Kölliker named them *dorsalia* – which shows that to him the posture of the vertebrate body still provided the natural reference system for all animals. He called the vertebrate appendages *lateralia* because they

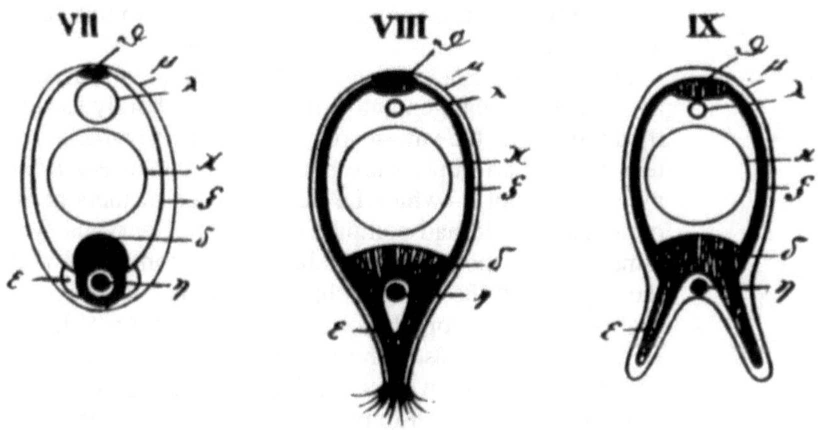

Fig. 3. Kölliker's comparison of arthropod legs with the dorsal plates in vertebrates (from Kölliker 1842, translations courtesy of K.Sander). *VII* A vertebrate cross-section upside-down. *IX* An insect cross-section in upright orientation. Between these, *VIII* represents a section through the first thoracic segment of the midge *Simulia*, showing median fusion of the ventral appendages (to be compared with the spinal arch of the vertebrate in *VII*). The Greek letters (arranged from top to bottom in the figure) indicate the following structures: ϑ dorsal muscles (insect) or sternum (vertebrate); μ skin; λ dorsal vessel (insect) or heart (vertebrate); κ stomach; ζ lateral muscles (insect) or ribs (vertebrate); δ ventral muscles (insect) or vertebral column (vertebrate); η nerve cord (insect) or spinal cord (vertebrate); ε leg muscles (insect) or spinal archs (vertebrate)

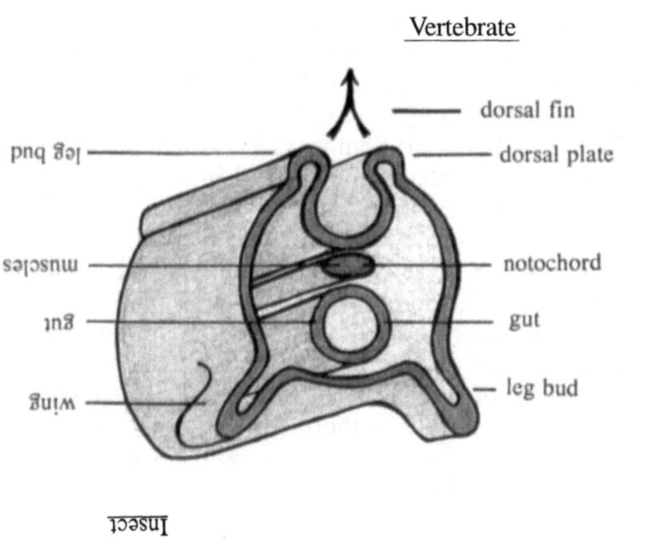

Fig. 4. Diagram to illustrate Kölliker's comparison (1842) of the origins of dorsoventral organization in a vertebrate (chick, *right*) and in an insect that has been turned upside down (*left*). The pair of upright folds (*top*) represent Kölliker's dorsal plates. To the modern reader these will clearly represent the neural folds, but Kölliker, like earlier embryologists, was not aware that their inner faces contribute to the neural tube; this is evident from his text and from Fig. 3 where the neural strand/tube is *flanked* by the derivatives of the (ectodermal!) dorsal plates. Note that in insects the hypothetical structures held equivalent to the dorsal plates do not fuse in the midline but each gives rise to the legs of one body side (the serial subdivision into individual legs and segments is not shown). The vertebrate legs, on the other hand, would have the insect wings as their counterpart. Insects, as compared to vertebrates, are thus evidently walking upside down

first appear halfway between the prospective dorsal and ventral sides. He added that arthropods, too, may form protrusions laterally, such as, for example, the gills in decapod crustaceans or the wings in insects. Switching finally to a very general viewpoint, Kölliker added that the divergent development of the dorsal plates in arthropods and vertebrates "has no other cause than that [well-known] gradual progression of all natural things."

Alternative interpretations by v.Baer (1828) and Zaddach (1854)

The view that the ventral side of insects corresponds to the dorsal side of vertebrates was not generally accepted even by those who acknowledged some of the observations on which this view was based. For example, the famous Karl Ernst von Baer (1792–1876, Professor at Königsberg and St. Petersburg) admitted that during *early* development the prospective ventral side of articulates is similar to the dorsal side of vertebrates and thus the respective nerve cords might correspond to one another. Yet von Baer noted that the functional ventral sides of arthropods and vertebrates display specific ventral structures such as mouth and (often) anus. He saw the basic difference in the fact that vertebrate embryos enclose their yolk from dorsal towards ventral while arthropods do so vice versa. Therefore one should state that arthropods develop from ventral to dorsal "rather than having them walk about on their backs" (von Baer 1828).

Gustav Zaddach (1817–1881, high school teacher and university reader at Königsberg University) reached similar conclusions when he compared early chick development with his observations on the development of the *Phryganidae* (Trichoptera). Zaddach noted that insects and vertebrates both form a bilaterally symmetric *Keimstreif* (germband or primitive streak) with a median groove, but he pointed out that in insects this groove (flanked by the two halves of the mesoderm) opens towards the yolk, while in vertebrates the primitive groove opens towards the surface of the germband (Fig. 5A). He conduded that in both forms the groove thus would mark the future dorsal surface of the germband. Zaddach therefore rejected the idea "that arthropods, as compared to vertebrates, turn their dorsal side towards the earth and their ventral side towards the sky" (Zaddach 1854).

To unify conceptually the further development in articulates and arthropods, Zaddach chose a fairly abstruse yet per se ingenious construction. He suggested that the insect body cavity corresponds to the amniotic cavity of vertebrates, not only in view of the median groove's position but also because the walls of both cavities (he thought) derive from advancing epithelial folds (Fig. 5). Adult insects thus lack the equivalent of the vertebrate

A

B

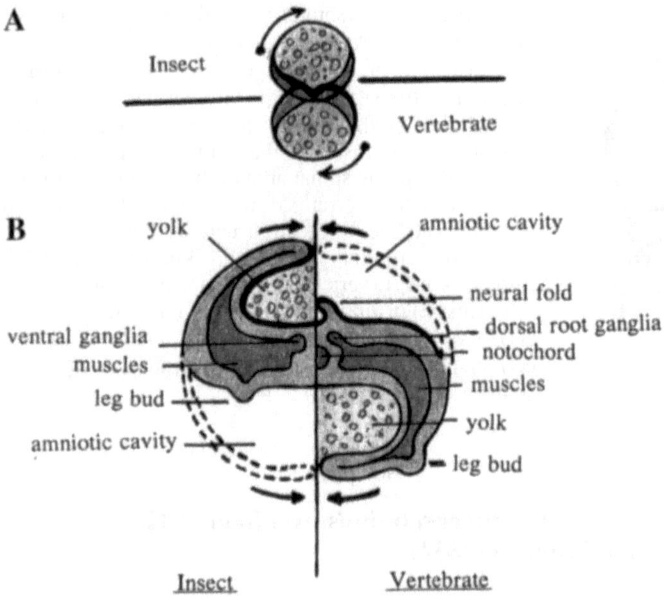

Fig. 5A,B. Diagram to illustrate Zaddach's idea (1854) of comparable body regions in an insect (*left*) and in a vertebrate (*right*). **A**, *upper half*: *Keimstreif* (germ band) in an insect. **A**, *lower half*: primitive streak in the chick. The median groove in the primitive streak and the groove between right and left halves of the germ band mesoderm, which he considered equivalent, indicated to Zaddach that the inner face of an insect's germ band corresponds to the outer face of the vertebrate primitive streak, which are both the dorsal boundaries of the respective structures (see text). *Arrows* indicate, respectively, dorsal closure in the insect and ventral incorporation of yolk in the chick. **B** Situation towards the end of (*left*) dorsal closure in an insect and (*right*) ventral incorporation of the yolk in a chick, with equivalent organs indicated according to Zaddach. For further details see text

body cavity because this cavity corresponds to their amniotic cavity (which they lose before hatching from the egg). Like Kölliker he stated that the vertebrate neural tube has no counterpart in insects, but (like von Baer) he took the paired row of insect ganglia and their segmental nerves to correspond to the paired row of vertebrate dorsal root ganglia and their spinal nerves (Zaddach 1854).

Leydig (1864) and Semper (1874): Comparative anatomy suggests a common ancestor for articulates and chordates

We now turn to concepts published after Darwin's *Origin of Species* (1859). Ernst Haeckel (1834–1909), Professor in Jena and a predominant figure in German biology, acknowledged in his influential *Generelle Morphologie* (1866) that one might utilize anatomical similarities between, for example, vertebrates and arthropods to elucidate their genealogical relationships, as well as the relationship between their dorsal versus ventral nerve cords. "Yet," he concluded "we hold these to be analogous and not homologous structures". Haeckel thought it not impossible that vertebrates originated independent of other animal phyla, and therefore recommended that one should refrain from trying to reconstruct any intermediary forms.

In contrast, Franz Leydig (1821–1908, comparative anatomist in Tübingen and Bonn) could "...barely escape the feeling that the various plans according to which vertebrates and arthropods are built, derive ultimately from a common root" – which he implied to be of metameric organization. He had compared the central nervous systems of arthropods and vertebrates and concluded that the supra- and the suboesophageal ganglia (i.e the cerebral mass) of arthropods correspond to the brain of higher animals, such "as if the anterior gut had pierced the vertebrate brain between its halves". Moreover, the ventral versus dorsal positions of the nerve cords in the Annulata (= articulates) and in the Vertebrata indicated to him that either "the dorsal nerve cord moved towards the lower regions of the body, far below the gut [...], or one had to assume – with Geoffroy St.-Hilaire – that [...] the entire organism inverted its orientation with respect to the substrate, so that an articulate would correspond to a vertebrate turned upside down" (Leydig 1864).

Ten years later Carl Semper (1832–1893, comparative anatomist in Würzburg), discovered in a shark's pronephros "true segmental organs", i.e metanephridia, which he considered homologous to the metanephridia in annelids (Semper 1874, 1878). Consequently, and contrary to the opinion of the dominant comparative anatomist Carl Gegenbaur (see below), he asserted that vertebrates are akin to annelids, and explicitly proposed resurrecting the "old concept of Etienne Geoffroy St.-Hilaire which Gegenbaur holds to be so thoroughly dispatched that he highlights his reference to same with an exclamation mark" (Semper 1874).

Anton Dohrn (1875): Vertebrates derive from annelid-like organisms that turned upside down during their evolution

For Anton Dohrn (1840–1904), founder of the famous "Stazione Zoologica" in Naples (1873), the origin of vertebrates always remained the central research topic (see Kühn 1950; for a detailed description of Dohrn's life, see Heuss 1940). Dohrn, too, had recourse to inverting the dorsoventral body axis. However, for him (and all subsequent authors) it was the vertebrates, not the articulates, that had turned upside down. This view was based, in addition to the similarity of the early segmented stages of articulate and vertebrate embryos, on Semper's discovery (just mentioned) of annelid-type segmental metanephridia in vertebrates. Dohrn thus derived vertebrates from annelid-like ancestors. This *annelid hypothesis* was welcomed by K.E. von Baer, among others; he found it "more appealing to derive vertebrates from annelids, which are at least composed of metameres, rather than from the quite unsegmented bag-like ascidians" – notwithstanding the fact that (like Haeckel) he did "not feel any need to derive the vertebrates from any other larger group".

Dohrn was the first to envision, for various organs, plausible and functional intermediary stages between an-

I.

Genealogischer Zusammenhang der Anneliden und Wirbelthiere.

Betrachtet man die ersten Stadien der Embryonalbildung eines Wirbelthiers, z. B. eines Knochenfisches, so kann man sich nur schwer dem Gedanken entziehen, dass man es mit einem in eine grosse Zahl von Segmenten gegliederten Geschöpf zu thun habe. Die sogenannten Urwirbel erscheinen als ebenso viel Segmente oder Metameren, und blickt man früh genug auf die Structur des Nervensystems, so gewahrt man deutlich 8—9 Segmente in der Gegend des vierten Ventrikels, der Rautengrube. Beide Bildungen sind principiell in nichts von der früh auftretenden Gliederung eines Insecten-Embryos zu unterscheiden. Im Speciellen freilich giebt es so reichliche Unterschiede, dass es eben bisher nirgends versucht wurde, diese Unterschiede als secundäre zu betrachten und über sie hinweg die Uebereinstimmung beider Embryonen zu betonen.

Was am meisten einer solchen Vergleichung im Wege stand, war die Anschauung, dass das Nervensystem der Gliederthiere am Bauche derselben gelegen sei, bei den Wirbelthieren aber am Rücken, bei jenen daher Bauchmark, bei diesen Rückenmark heisst. Hätte man diese Bezeichnung Rücken und Bauch nicht gehabt, so wäre die Vergleichung schon wesentlich leichter gewesen und man würde vor Allem bei den Embryonen beider Thierkreise gleich von vornherein auf eine grosse Aehnlichkeit gestossen sein: auf die Lagerung des Nervensystems an der convexen Seite des Embryoleibes. Mit

Dohrn, Functionswechsel. 1

Fig. 6. Opening paragraphs of Dohrn's 1875 booklet discussing the genealogical connection between annelids and vertebrates. Dohrn here explains that (1) the differences between insect and vertebrate embryos hitherto held to be important might be secondary while no one had tried yet to emphasize the many similarities between these embryos, and (2) that without the terms back and belly the position of the nerve cord on the embryo's convex side, common to both, would have been immediately obvious

cestral annelids and chordates, and thereby introduced the "principle of change of functions". In his booklet *Der Ursprung der Wirbelthiere und das Princip des Functionswechsels* (1875; see Fig. 6), dedicated to K.E.von Baer, Dohrn also discussed – and may have solved in part, – the most serious problems that arose from the idea of dorsoventral inversion during chordate evolution.

One major problem was that in articulates the foregut passes through the brain (between the supra- and the suboesophageal ganglia) and opens into a mouth at the ventral side, while in vertebrates the foregut reaches the mouth without penetrating the brain (see Fig. 1). From this, Dohrn concluded that the mouth of vertebrates is a new opening that formed on the new ventral (i.e formerly dorsal) side of the organism. He proposed that the new mouth arose by fusion of the anteriormost pair of gill slits, because (1) the vertebrate mouth breaks through rather late during development (i.e. well after what is now called the "phylotypic stage") while in other phyla the mouth forms very early, usually as a derivative of the blastopore; (2) the vertebrate mouth is accompanied by a pair of branchial arches, as are the gill slits; and (3) it develops simultaneously with, but in front of, the anteriormost pair of definitive gill slits, i.e. far more posteriorly than its later position in the body would predict.

But where did the segmentally arranged gill slits come from? After all, they exist neither in annelids nor in arthropods. Dohrn proposed that the vertebrate gill slits derive from segmental nephridial tubules that fused to the anterior gut and thereby connected the lumen of the gut – via the nephropore – to the surrounding water. These tiny lateral openings could originally have served to extrude water that was ingested through the old mouth. It then "appears easy to envision how some of the gill slits began to ingest water into the gut. If now the slits [...] widened until they fused at the dorsal midline, a rival mouth arose which then gradually perfected itself so that the complete function of food intake was transferred upon it." – Dohrn conceded, however, that the early function of the gill slits was not quite clear.

With his "tracing back of the vertebrate segmental gill slits to former segmental nephridial openings, with the new vertebrate mouth derived from a fused pair of former gill slits, and with the region of the medulla oblongata [or the infundibulum] as the former passageway in between the circumpharyngeal connectives", Dohrn felt that vertebrates were satisfactorily linked to an annelid-like organism. Interestingly, and perhaps guided by his earlier work on evolutionary degeneration in parasitic crustaceans, he derived *Amphioxus* and the unsegmented ascidians via degeneration from fish-like vertebrate ancestors, as did von Baer (1873 and below).

Dohrn's ideas contracted the wrath of his old teacher Haeckel (1875) and of the famous comparative anatomist Carl Gegenbauer (1826–1903). For example, Gegenbaur (1876) quoted as a "striking example... of unscientific comparison the well-known equation of the ventral nerve cord of invertebrates with the dorsal spinal cord of vertebrates." However, Alfred Kühn (1950) in his impartial yet sympathetic description of Dohrn's scientific work reported that "... a person who [like Haeckel] links very divergent animal groups in a phylogenetic tree connecting them by simple lines is far less exposed to criticism than he who [like Dohrn] undertakes to trace the transformation of their specific organization step by step."

Dohrn himself considered his 1875 book as a "preliminary chalk sketch to an *al fresco* painting". He began to search for indications of primitive evolutionary stages in the early embryogenesis of various lower vertebrates. His series *Studien zur Urgeschichte des Wirbelthierkörpers* (Studies on the developmental history of the vertebrate body) include, for example, investigations on the ontogenesis of the vertebrate mouth and gill slits and of the pituitary (thought to indicate the location of the old mouth), on the development of spinal nerves, branchial arches and somites in the posterior head region (Dohrn 1881–1907).

It may be interesting to compare Dohrn's approach with that of his most influential critic, Gegenbaur, because the two of them personify the differing strategies for reconstructing the origins of a specific structure during evolution. Gegenbaur's approach, which one might call "phylogenetic morphology", is based on comparative anatomy and palaeontological evidence, and looks for similarities in adult structures in order to deduce homologies, while "phylogenetic embryology" as practised by Dohrn hopes to find – by comparing the specific modes of ontogeny – indications of how this structure might have emerged during evolution.

Impact of Dohrn's ideas

In spite of their being ostracized by Gegenbaur and Haeckel, Dohrn's ideas had some influence on other biologists. For example, Hatscheck (1878) stated that the similar morphology and development of annelids and vertebrates allows one to assume a segmented common ancestor, and so did Beard (1889) and Minot (1897). The homology of the vertebrate mouth with an anterior pair of gill slits was discussed by van Wijhe (1893) and Beard (1888). Even the great Richard Owen (1883) assumed homology between the nervous systems of articulates and vertebrates, stating like Dohrn that the vertebrate pituitary indicates the location of the old mouth. Dohrn in turn felt encouraged by the findings of others, even if they did not believe in his annelid hypothesis. An example is the series of mesodermal cavities, one pair per branchial arch, in the head region of embryonic selachians, described in 1877 by Frank Balfour (1851–1882, comparative embryologist in Cambridge), a frequent visitor to Naples and, together with Ernst Abbe, Dohrn's closest friend. Balfour rejected the annelid hypothesis, but Dohrn felt that these mesodermal cavities could be considered homologous to the segmental coelomic cavities in annelids (compare Fig. 7 and Fig. 8).

Adolf Naef (1926): Ontogenesis in the hypothetical common ancestor, and its modification towards chordate development

During the first decades of the twentieth century, phylogenetic speculations based on embryogenesis were generally despised. Nonetheless, some 50 years after Anton Dohrn, the Swiss zoologist Adolf Naef (1883–1949) endeavoured to revive Dohrn's "annelid hypothesis" of chordate ancestry – which he considered a truism beyond discussion. Naef aspired to work out the "archetypal" ontogenesis of an "idealized" primeval coelomate of bilateral and segmented organization, suitable as a common ancestor of both annelids and chordates (Fig. 7). From this archetypal ontogenesis he then derived the mode of embryogenesis in an idealized ancestral chordate (Fig. 8). As for the bodily "revolution" during chordate evolution, Naef proposed that the worm-like chordate ancestor lived buried in sand or mud where constant dorsoventral orientation was of no importance, as is the case in some living hemichordata, e.g. *Balanoglossus* (Naef 1933).

Note that the terms "archetypal" and "idealized" do not mean that Naef returned to the principles of idealistic morphology as claimed by Mayr (1982). Rather, Naef's aim (like that of Dohrn) was to reconstruct, on the basis of comparative embryology, the specific course of possible ontogenetic changes in the evolution of chordates from an annelid-like ancestor. But thereby, in contrast to Dohrn, he considered *Amphioxus* and the ascidians to be truly pri-

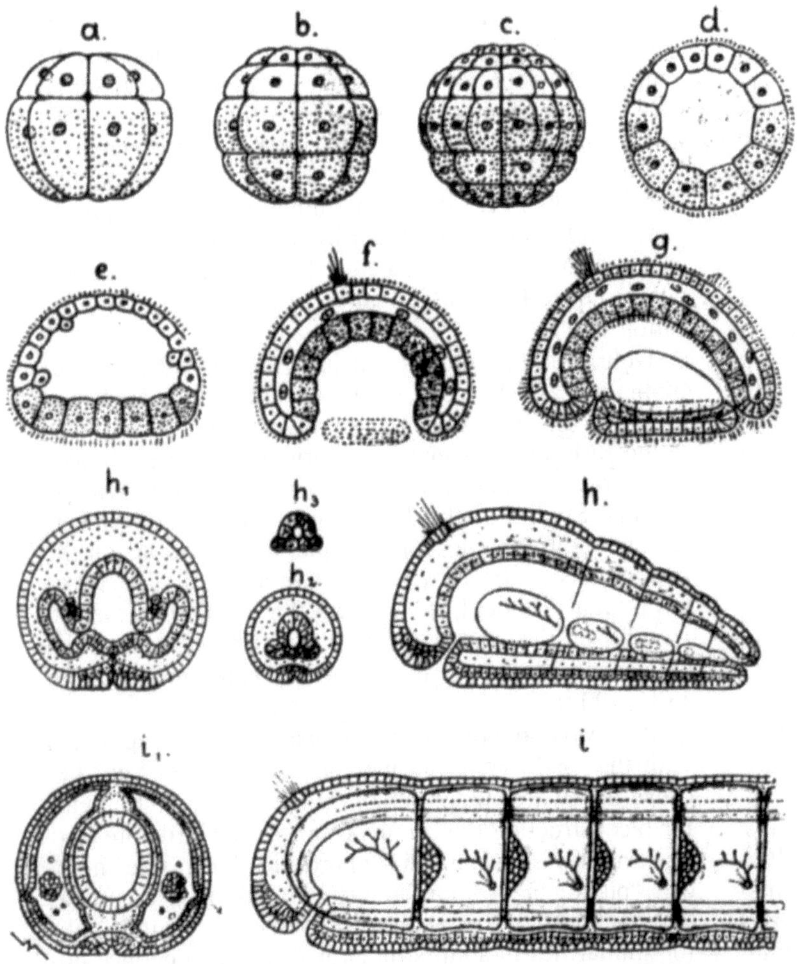

Fig. 7. Archetypal ontogenesis of an idealized ancestral segmented coelomate organism according to Naef (1926a). For explanation, see text

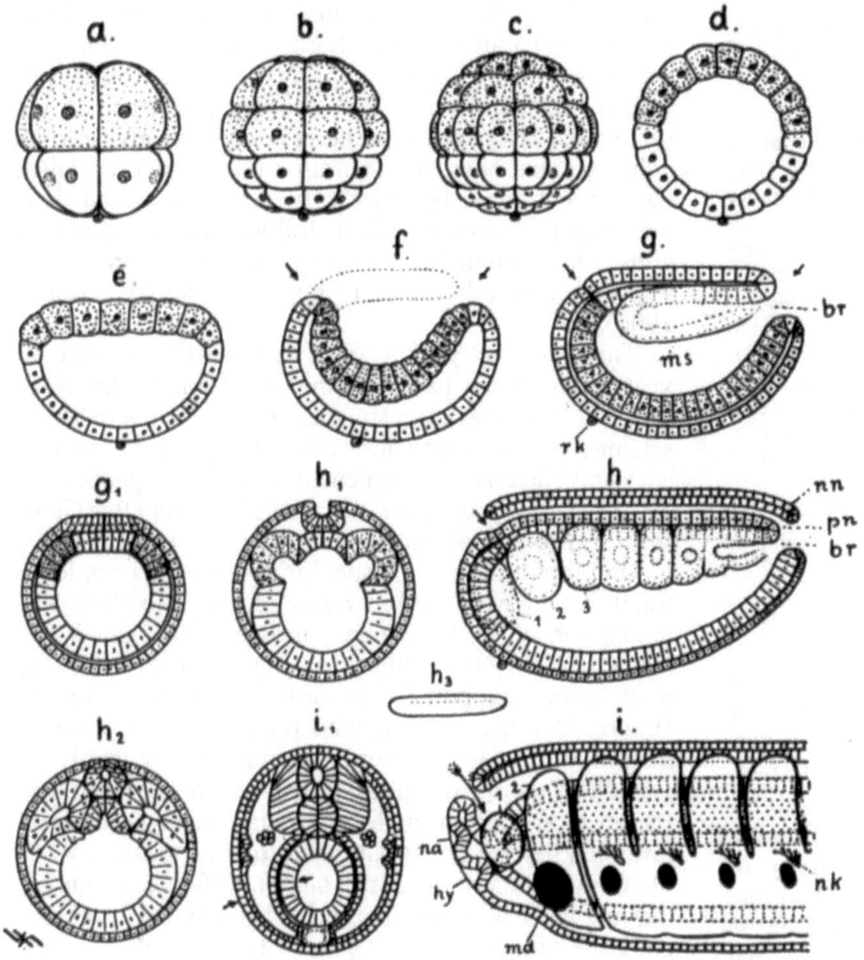

Fig. 8. Archetypal ontogenesis of an idealized ancestral chordate organism according to Naef (1926a). Note that, as compared to Fig. 7, the body's dorsoventral axis has been reversed (as in Fig. 1), beginning with an inversion of the animal-vegetal egg axis in **a**. The arrows in *f* and *g* indicate the location, in the ancestral coelomate (Fig. 7), of anus (right) and mouth (left, still marked – in Naefs view – by a slight depression in the late gastrula of *Triturus* and persisting in the pituitary region (*hy* and *i*)). Abbreviations *br* remnant of blastopore, *hy* ectodermal pituitary rudiment, *md* mouth, *ms* mesoblast, *na* olfactory pit, *nk* nephridial duct, *nn* median fusion line of neural folds, *pn* canalis neurentericus, *rk* polar body. For further explanation see text

meval chordates and not the products of degeneration from fish-like vertebrates.

Basic features of the primeval coelomate's early development according to Naef were a defined animal-vegetal egg axis (Fig. 7a), the immigration of some ectodermal cells into the blastocoel to form a mesenchyme (Fig. 7e), and gastrulation by invagination of the vegetal hemisphere. During gastrulation the animal pole (marked by an apical tuft), was to move slightly towards the prospective anterior body region, thus yielding a bilateral organism with anterior, posterior, dorsal and ventral sides (Fig. 7f). Median fusion of the lateral but not the anterior or posterior lips of the blastopore would then leave open mouth and anus (Fig. 7g). The central nervous system was to develop (as in recent polychaete annelids) along the fused former margins of the slit-like blastopore (Fig. 7, h and h1), thus yielding a ventral nerve cord anteriorly expanded into a brain.

For this common ancestor of articulates and chordates Naef further postulated the existence of a posterior growth zone where the elongating archenteron sequentially forms lateral pockets that later pinch off to produce a series of coelomic cavities on either side (Fig. 7, h and h1). – Coelom formation (from the 4d cell) in today's annelids derives, according to Naef, from this enterocoelous mode of mesoderm formation: progressively fewer cells of the posterior stretch of archenteron tended to pinch off, and this occurred increasingly early in embryogenesis until a single cleavage cell remained for that task. This then multiplied and its progeny subdivided into small cell groups that formed coelomic cavities similar to those seen today after enterocoelous mesoderm formation in lower chordates (Fig. 7, h1–h3). – In accordance with Semper's findings (see above), this idealized coelomate embryo then formed one nephridium per coelomic cavity. The nephridial ducts were envisioned to associate with several solenocytes (Fig. 7h,i), as is the case today in archiannelids which Naef thought came closest to the common ancestor.

Starting from this hypothetical common ancestor of articulates and chordates, and embodying some specific features of early *Amphioxus* development meanwhile described by Hatschek (1881) and others, Naef then reconstructed the main features of an ancestral chordate's ontogenesis (Fig. 8). He thereby inverted – for purely didactic reasons – the animal-vegetal egg axis (Fig. 8a) to match the inverted adult posture (compare Fig. 7i and Fig. 8i). The blastopore would then close around the upper (= vegetal) pole. In contrast to closure of the blastopore by advancing lateral lips as in the ancestor (Fig. 7f), the lateral lips in the primeval chordate are assumed to fuse zipper-fashion in anteroposterior progression, until a single opening remains at the posterior, to become the anus (Fig. 8f,g). Naef, like Dohrn in his early papers, proposed that the brain region formerly pierced by the ancestor's foregut corresponds to the infundibulum of the vertebrates.

In a chordate embryo as seen in Fig. 8g, the posterior growth zone would already be fully active. The dorsolateral region of the primary gut successively forms the future coelomic cavities (Fig. 8 h, h1 and h2), and its dorsal region below the neural plate differentiates into the notochord (Fig. 8, g1, h1, h2 and i). The neural plate folds up to form the neural tube (Fig. 8g1, h1, h2 and i). The ventral sectors of the coelomic cavities then fuse to form the lateral plates, and the segmental metanephridia (with associated solenocytes) develop between them and the metameric dorsal somites (Fig. 8i and i1). – The gill slits (black in Fig. 8i) originally evolved, according to Naef, in strict correlation with the nephridial tubules. He postulated this because he believed to have seen that in *Amphioxus* the nephridial tubules become confluent with the gut during ontogenesis (Naef 1939), thus connecting the gut lumen to the surrounding water, as proposed by Dohrn in 1875. – Naef considered the two anteriormost pairs of coelomic cavities (Fig. 8i) to represent the premandibular and the mandibular somites, respectively. He thought that the gill slits at the mandibular somite (2) later fused medioventrally, thereby forming the definite mouth of vertebrates (md in Fig. 8i). This mode was to hold only for the vertebrate ancestor while the well-known unilateral mouth formation in *Amphioxus* was considered proof of the early separation of Vertebrata and Acrania.

Today, we associate the name of Naef with his other main field of interest, the ontogeny and evolution of cephalopods, whereas he failed to gain recognition for his attempts at elucidating large-scale phylogeny step by step, starting from Dohrn's "annelid hypothesis" of vertebrate evolution and the concomitant dorsoventral inversion of the chordate ancestor. In accordance with his being "but a morphologist", Naef was never offered a permanent academic position in his home country or elsewhere in the German-speaking countries. Therefore he had to accept a chair at Cairo in Egypt where, faced with various hardships and isolated from international science, he continued his work until his death shortly after Second World War.

Conclusion

Geoffroy St.-Hilaire's idea (1822) that vertebrate anatomy resembles the anatomy of a dorsoventrally inverted arthropod has re-emerged about every 50 years, with increasing recourse to embryological and, more recently, molecular data (Arendt and Nübler-Jung 1994). In the post-Darwinian era it was Anton Dohrn who, influenced by Darwin and by Semper's findings of annelid-like metanephridia in vertebrates, in 1875 put forward the idea that chordates derive from annelid-like ancestors that had inverted their dorsoventral body axis.

Dohrn's lifelong quest to find supporting evidence for this "annelid hypothesis" remained unconvincing even in his own judgement, possibly because he refused to consider lower chordates (*Amphioxus*) as possible intermediaries between annelids and vertebrates. Kühn (1950) noted that Dohrn, with his annelid hypothesis, underwent an experience similar to Goethe's with his vertebral theory of the skull: "As a whole it can be pronounced but not proven, in detail it well could be presented but one will vainly strive to get it rounded off and complete" (Goethe 1830b). Consequently, Dohrn grew more and more hesitant about his own speculations but, in general, never gave up the annelid hypothesis.

Around the turn of the century other, more experimental disciplines were to supersede the hitherto flourishing comparative embryology, which had been triggered and inspired by the impact of Darwin's work. Yet, Dohrn vehemently defended this discipline, and in his second but last publication he gave specific advice to its adherents on how "to restrain the arrogant criticism handed out to it by *Entwicklungsmechanik* and *Entwicklungsphysiologie* which follow entirely different aims, and to strengthen its independent and fundamental importance as a discipline of historical biology" (Dohrn 1906).

This was exactly what Adolf Naef tried to do some 50 years after Dohrn had published his "annelid hypothesis", but there was no resonance in spite of his coherent schemes of how chordates could have evolved from an ancestral bilateral and segmented coelomate organism.

Another 50 years later – that is, 150 years after Geoffroy St.-Hilaire had maintained that dorsal in chordates is ventral in arthropods – specific genes have been identified that are expressed in ventral domains in *Drosophila* and dorsally in vertebrates, or vice versa (Arendt and Nübler-Jung 1994). Molecular genetics thus seems to provide some support for the basic ideas of Geoffroy St.-Hilaire, Dohrn and Naef, who propagated the *unité de plan* of articulates and chordates, implying the bodily inversion of the dorsoventral axis in one of these groups. However, supporting molecular data cannot prove axial inversion beyond doubt, and therefore the closing words of Anton Dohrn in his 24th *Studie* (1906) are as true today as when he wrote them: "What odysseys this hypothesis will yet have to endure I shall here not dare even to hint. That it will have to endure some, I feel will be unavoidable but also no particular misfortune. After all, even the most vehement antagonism of opinions is but an advantage for the final outcome since it will instigate ever new attempts to expand the realm of the factual and to deepen that of critical understanding."

We are greatly indebted to K. Sander for improving this manuscript, for drawing our attention to the work of Zaddach and Kölliker, and for his constant enthusiasm and encouragement, also while we were working out ideas we have published elsewhere. In addition, we thank R. Spoerle for drawing our attention to the work of Adolf Naef and J.W. van Wihje, C. Groeben (Naples) for sending a copy of Dohrn's 1875 book, S. von Boletzky (Banyuls-sur-mer) for information on the life of A. Naef, and the staff of the University Library in Tübingen for their friendly assistance.

References

Appel TA (1987) The Cuvier-Geoffroy debate. French Biology in the decades before Darwin. Oxford University Press, New York

Arendt D, Nübler-Jung K (1994) Inversion of the dorsoventral axis? Nature 371:86

Baer KE von (1828) Ueber Entwickelungsgeschichte der Thiere. Beobachtungen und Reflexionen, 1. Theil. Königsberg

Baer KE von (1873) Entwickelt sich die Larve der einfachen Ascidien in der ersten Zeit nach dem Typ der Wirbelthiere? Mem Acad Imp Sc Pétersbourg (7 Série) XIX (8):1–35

Balfour F (1877) The development of elasmobranch fishes. J Anat Phys Lond 11 (cited after Kühn 1950)

Bateson B (1886) The ancestry of chordates. Q J Microsc Sci XXVI, London (cited after Lönnberg 1902)

Beard J (1888) The old mouth and the new. A study in vertebrate morphology. Anat Anz 3:15–24

Beard J (1889) Some annelidan affinities in the ontogeny of the vertebrate system. Q J Microsc Sci Vol XXXiX, No 1002, p 259, London (cited after Lönnberg 1902)

Cuvier G (1817) Le règne animal. Paris (cited after Jahn et al. 1985)

Darwin C (1859) On the origin of species by means of natural selection. Murray, London

Dohrn A (1875) Der Ursprung der Wirbelthiere und das Princip des Functionswechsels. Genealogische Skizzen. Verlag von Wilhelm Engelmann, Leipzig. English translation: Ghiselin MT (1994) The origin of vertebrates and the principle of succession of functions. Genealogical sketches by Anton Dohrn. 1875. An English translation from the German, introduction and bibliography. Hist Phil Life Sci 16:5–98

Dohrn A (1881–1907) Studien zur Urgeschichte des Wirbelthierkörpers 1–25. Mitt Zool Stat Neapel, Vols. 3–18

Garstang W (1894) Preliminary note on a new theory of the phylogeny of the chordates. Zool Anz 27:122–125

Gaskell W H (1898) On the origin of vertebrates deduced from the study of ammocoeta. I. The origin of the brain. J Anat Phys Lond XXXII:513–553

Gegenbaur C (1876) Die Stellung und die Bedeutung der Morphologie. Morph Jahrb Anat Entwgesch 1:1–19

Geoffroy St.-Hilaire E (1822) Considérations générales sur la vertèbre. Mém Mus Hist Nat 9:89–119

Goethe JW von (1830a) Commentary on "Principes de Philosophie Zoologique par Geoffroy Saint-Hilaire". Goethes Werke, vollständige Ausgabe letzter Hand (1833) 50:201–250. Cotta, Stuttgart

Goethe JW von (1830b) Ueber den Zwischenkiefer des Menschen und der Tiere. Verh Kaiserl Leopoldin-Carol Akad Naturforsch 15 (1830) (cited after Kühn 1950)

Goette A (1895) Über den Ursprung der Wirbelthiere. Verh Dtsch Zool Ges 5. Jahresversammlung zu Strassburg. (cited after Lönnberg 1902)

Gutmann W F (1967) Die Entstehung des Coeloms und seine phylogenetische Abwandlung im Deuterostomier-Stamm. Zool Anz 179:109–131

Gutmann W F (1976) Phylogenetic reconstruction: theory, methodology and application to chordate evolution. In: Hecht MK, Goody PC, Hecht BM (eds) Major patterns in vertebrate evolution. Plenum Press, New York, pp 645–669

Haeckel E (1866) Generelle Morphologie der Organismen, vol 2. Allgemeine Entwicklungsgeschichte. Berlin 1866.

Haeckel E (1875) Ziele und Wege der heutigen Entwicklungsgeschichte. Jena Z Naturwiss, Suppl 10

Hall BK (1992) Evolutionary developmental biology. Chapman and Hall, London.

Hassenstein B (1958) Prinzipien der Vergleichenden Anatomie bei Geoffroy Saint-Hilaire, Cuvier and Goethe. In: Goethe et l' Ésprit Francais. Société d'Éditions: Les Belles Lettres. Paris

Hatschek B (1878) Studien über Entwicklungsgeschichte der Anneliden. Arb Zool Inst Wien 11:1–29

Hatschek B (1881) Studien über Entwicklung des *Amphioxus*. Arb Zool Inst Wien 4:1–88

Heuss T (1940) Anton Dohrn in Neapel. Atlantis- Verlag, Berlin. English translation: Anton Dohrn. A life for science. (1991) Springer, Berlin Heidelberg New York.

Jahn I, Löther R, Senglaub K (1985) Geschichte der Biologie. Fischer, Jena

Note added in proof: The claim that the nerve cords of vertebrates and articulates are homologous was also raised by H. S. Greenough (1892; see the Essay on the stereomicroscope).

Jefferies RPS (1991) Two types of bilateral symmetry in the metazoa: chordate and bilaterian. In: Bock GR, Marsh J (eds) Biological asymmetry and handedness. Ciba Foundation Symposium 162. Wiley, Chichester, pp 49–127

Kaestner A (1993) Lehrbuch der speziellen Zoologie, vol I. Grell KG, Gruner H-E, Kilian EF (eds). Fischer, Stuttgart.

Kölliker A (1842) Observationes de prima insectorum genesi adiecta articulatorum evolutionis cum vertebratorum comparatione. Inaugural dissertation. Meyer und Zeller, Zürich. (For German translation, see Wieland 1993)

Kühn A (1950) Anton Dohrn und die Zoologie seiner Zeit. Pubbl Staz Zool Napoli [Suppl] 1950. Editione della Stazione

Leydig F (1864) Vom Bau des thierischen Körpers, vol 1. Laupp and Siebeck, Tübingen (cited after Kühn 1950)

Lönnberg E (1902) Über den Ursprung der Vertebraten. In: Bronn's Klassen und Ordnungen des Tier-Reiches, vol 6: 83–98. Wintersche Verlagsbuchhandl, Leipzig

Mayr E (1982) The growth of biological thought. Belknap Press/Harvard Univ Press, Cambridge, Mass.

Minot CS (1897) La détermination des ancètres des vertébrés. Arch Zool Exp Gen (3e série): 417–436

Naef A (1926a) Notizen zur Morphologie und Stammesgeschichte der Wirbeltiere. 7. Das Verhältnis der Chordaten zu niederen Tierformen und der typische Verlauf ihrer frühen Entwicklung. Biol Zentralbl 46:39–50

Naef A (1926b) Notizen zur Morphologie und Stammesgeschichte der Wirbeltiere. Pubbl Staz Zool Napoli, vol VII, 299–333

Naef A (1933) Vorstufen der Menschwerdung. Fischer, Jena

Naef A (1939) Die Entwicklung der Kiementaschen und Nephridien beim larvalen und metamorphosierenden Lanzettfischchen. Zool Jahrb Anat 65:469–516

Owen R (1883) Essays on the conario-hypophysial tract and on the aspects of the body in vertebrate and invertebrate animals. London (cited after Naef 1926b)

Patten W (1890) On the origin of vertebrates from arachnids. Q J Microsc Sci, vol 31, London (cited after Lönnberg 1902)

Rathke H (1829) Ueber die Bildung und Entwickelung des Flußkrebses. Verlag Leopold Voss, Leipzig

Remane A (1967) Die Geschichte der Tiere. In: Heberer G (ed) Die Evolution der Organismen, vol I Fischer, Stuttgart, pp 589–677

Romer AS, Parsons TS (1977) The vertebrate body. Saunders, Philadelphia

Semper C (1874) Die Verwandtschaftsbeziehungen der gegliederten Tiere. Arb Zool Inst Würzburg 2 (1874) (cited after Kühn 1950)

Semper C (1878) Sind die Segmentalorgane der Anneliden homolog mit denen der Wirbeltiere? Eine Erwiderung an Herrn Dr. Fürbringer. Morphol Jahrb 4:322–327

Slack JMW, Holland PHW, Graham CF (1993) The Zootype and the phylotypic stage. Nature 361:490–492

Starck D (1978) Vergleichende Anatomie der Wirbeltiere, vol 1. Springer, Berlin Heidelberg New York

Tort P (1983) La querelle des analogues. Éditions d'Aujourd'hui, Plan de la Tour (Var), France

Wieland M (1993) Frühe Untersuchungen über die Insekten-Embryogenese. Ein Beitrag zur Geschichte der Entwicklungsbiologie. Staatsexamensarbeit an der Fakultät für Biologie, Universität Freiburg i. Br.

Wijhe J van (1893) Ueber Amphioxus. Anat Anz 8:152–172

Zaddach G (1854) Untersuchungen über die Entwickelung und den Bau der Gliederthiere I. Heft. Die Entwicklung des *Phryganiden*-Eies. Verlag von Georg Reimer, Berlin

Wilhelm Roux
(1850–1924)

CENTENNIAL ESSAY

S.J. Counce

Archives for Developmental Mechanics W. Roux, Editor (1894–1924)

Introduction

This year, 1994, marks the centennial anniversary of *Roux's Archives for Developmental Biology*. On 16 October, 1894, Number 1, Volume 1 of the *Archiv for Entwicklungsmechanik*, the first journal devoted to experimental studies of development, was issued. For the next 30 years, its founder, Wilhelm Roux, was its sole editor. During this period, this "new branch" of biology (Roux 1905), the causal analysis of development, would grow into a vigorous science, and the influential *Archives* (Gascoigne 1985) would prosper as well. Roux's philosophy as editor was defined in the first paper in the journal (Roux 1894a). [A translation by W.M. Wheeler is reprinted in Maienschein 1986]. I will examine how this philosophy was practiced during the ensuing three decades. Areas of research that attracted the interest of early adherents of analytical embryology as reflected in the articles published in the *Archives* will be briefly explored. I will consider some seminal papers and ideas that first appeared in the *Archives*, as well as some of its contributors who are familiar figures to present day developmental biologists.

Wilhelm Roux and developmental biology

Roux is generally considered the founding father of experimental embryology, not because he was the first to use the experimental approach, but because he defined, in great detail, a program for *analytical embryology* or *developmental mechanics* (Sander 1991a). He preferred the latter designation as stressing that the ultimate goal of this approach was to explain developmental events in terms of the laws of physics and chemistry (Roux 1894a, 1895a (Vol. 2, 1-94), 1905; Churchill 1973, 1975; Maienschein 1991a; Sander 1991a). Roux was a superb publicist for his program: he took advantage of every available opportunity, every scientifc platform, to propagandize, to proselytize for this "new science." And the *Archives* provided a major platform.

Roux's editorial policies

Roux recognized that science was international in scope. To this end, he obtained endorsements from over one hundred leading European and American scientists (Roux 1894a) before he began publishing the *Archives*. From the first, the journal provided an open forum for experimental biologists from many nations. (For an overview of the development of experimental embryology in Europe and the United States see Horder et al. 1986, pp. 61–71, 109–112, 179–181; Ridley 1986.)

Roux (1894a) had a modern view of the fields of study that could contribute to an understanding of the complexities of development, and his breadth of view will seem familiar and appropriate to present-day developmental biologists. He spelled out in detail those areas of research that would be published in the *Archives*. He considered that important insights could be gained from any study relating structure, form and function, whether information was arrived at by experimentation or deduced from study of normal development, pathological specimens, anatomical anomalies, or "naturally occurring experiments." Theoretical papers were welcomed, as were critical reviews of the literature. His introductory section on the scope of the new journal concluded: "Because Developmental Mechanics values all methods that give rise to causal knowledge, and utilizes all biological disciplines for its purposes, its area of investigation encompasses all living organisms from the lowest protistans to the highest animals and plants."

Roux ended his introduction with what was to be a major touchstone in his own writings and in editorial practices: "It is an old saying that the meaning of Truth is born out of conflict. To best serve the Truth [we] will provide space for the most diverse concepts if they are supported by fact" (1894a, p. 36). However, he cautioned authors to use a careful tone in their disputes. "The search for Truth is not enhanced by personal enmity. [But] the more lively the battle is waged between different ideas, the more quickly we will approach our distant goal." (See Sander 1991c, p. 299 for a slightly different translation.) This splendid sentiment, however, was to be honored far more in the breach than in the observance by many contributors to the *Archives.*

Roux never hesitated to use his editorial prerogatives. He became notorious for editorial insertions in articles, in book reviews, even in brief literature listings. These were almost invariably to cite his own works on the same subject, or to twit "non-believers." He developed an almost pathological interest in establishing priority, real or imagined (Churchill 1975; Maienschein 1991a; Sander 1991a; see also Goldschmidt 1956, p. 73). In spite of this intervention, scientists continued to publish in the *Archives.* A major reason for this, perhaps, is found in a letter from Boveri to Spemann, which was brought to my attention by Klaus Sander. Written in response to Spemann's stated intention of no longer publishing in the *Archives* because of editorial intrusion, Boveri replied: "I must say that I am not in complete agreement with your decision. By this I mean one can and should separate Person and Thing (as with Driesch). It is without doubt a very significant advantage if all related works can be found as far as possible in one place." To his credit, Roux did not hesitate to publish the papers of those who opposed his views.

Even at the end of his active participation in the *Archives,* which fell off toward the end of his life, he maintained old habits. He appended an extensive *"Comments of the Editor"* to a paper by Giglio-Tos (1923), dealing with the author's omission of early papers of Roux. A reference in 1924 to Roux in a footnote in a subsequent Giglio-Tos paper suggests that at least in early 1924, Roux was still performing editorial tasks. For this reason, all volumes through 104 (1924) are included as part of Roux's editorial reign. From 1895 through 1918, the *Archives* were published by Engelmann in Leipzig; beginning with Vol. 45 (1919) Julius Springer, in Berlin, became the publisher.

During the three decades Roux was editor, no standard form for articles was required. He did request that contributors provide a summary of major conclusions at the end of each article. Unfortunately, he did not insist on this, for it would make reading of many of these complicated papers much easier. Before World War I, articles were published in Italian, French and English as well as German. Beginning in 1903, German summaries were provided for all non-German language papers. By 1920; with the exception of rare papers in English, all papers were in German.

Roux was assisted by faculty members at Halle and other institutions with literature reviews. But in so far as one can judge from easily detected "editorial insertions," Roux did not rely on others to review manuscripts. Nor did he place limitations on the length of papers; almost a third were 50 or more pages long. This makes the average publication period of 4–5 months between submission and publication of papers, regardless of length, language, or country of origin, remarkable.

In addition to the numerous articles and books he reviewed, often at considerable length, Roux frequently contributed papers to the *Archives,* continued to publish elsewhere, and edited other occasional publications devoted to developmental mechanics. At the same time, he headed the Anatomy Institute at Halle and was active in many professional societies (Barfurth 1920, 1925; Roux 1923). Roux's accomplishments as editor of the *Archives* must be considered quite phenomenal.

Roux was sole editor of the *Archives* for most of its first 30 years. Only in 1923, when it was merged with the *"Archiv für mikroskopische Anatomie und Entwicklungsgeschichte,"* was "with the assistance of H. Braus and H. Spemann" added to his name in the masthead. With his death this brief union ended, and in 1925 the journal became *"Wilhelm Roux' Archiv für Entwicklungsmechanik."*

Contributors

Americans were among the first and most prolific contributors to the *Archives* during its first two decades. That American scientists were major contributors to the *Archives* in its early years is not surprising. In the late nineteenth century, several universities in the United States modelled their graduate programs on the German university system. Americans also regularly travelled to Germany for study and to earn advanced degrees. The Zoological Station at Naples provided another institution where American scientists came in regular contact with their German colleagues (Allen 1978; Mayr 1986).

There were a surprising number of women among these American contributors. Most were students of E.B. Wilson and his successor T.H. Morgan at Bryn Mawr, one of the few institutions in America offering the doctorate to women at this period. Among Morgan's most distinguished Bryn Mawr students were Nettie Stevens, Helen Dean King and Florence Peebles (Rossiter 1982), all of whom published in the *Archives.* Three Bryn Mawr women, Nettie Stevens, Alice Boring and Mary Hogue, worked at the Naples Station and with Boveri at Würzburg. It is probable that during her two visits to Germany, Stevens became acquainted with Roux. She is the only woman and the only American honored with an obituary in the Archives (Roux 1912); he (1920) also listed her as one of his deceased colleagues.

Barbara Heffner, from Boveri's department, was the first German woman to publish in the *Archives* (1908). Thereafter, contributions from European women appeared with some frequency, especially from institutions in the Austro-Hungarian Empire. The most prolific European woman contributor was Leonore Brecher from Vienna, who published many papers on the biology and biochemistry of lepidopteran pigmentation. Rhoda Erdmann, the distinguished German scientist who is best known as a pioneer in tissue culture studies, began her scientific career in protozoology. She contributed reviews of the recent literature in protozoology (1913–1914) to the *Archives* and also a paper (in 1920) on protozoological studies she had done while a guest researcher at Yale University. On her return from America, she was appointed a docent in protozoology in the Philosophy faculty at Berlin, but shortly thereafter was named head of the Experimental Cytology Section of the University Cancer Research Institute (Roux 1921). Her first paper on tissue culture, an analysis of the behavior of heart valve tissues of reptiles and mammals in culture, appeared in the *Archives* in 1921.

Major European contributors were from academic institutions in Germany and the Austro-Hungarian Empire (and after World War I, the "new nations" of Eastern Europe, and Russia). Swiss, Italian, Dutch, and Scandinavian scientists were also relatively frequent contributors. Papers by French, British and Belgian scientists were rare.

Topics and papers – the first three decades

While Roux was editor of the *Archives* over 1200 papers appeared, dealing with a wide range of subjects. In content, these reflect the breadth of view expressed by Roux in the first paper in the *Archives*. In a short essay, however, only selected topics can be considered.

As indicated earlier, Roux believed that genuine differences of intellectual opinion could lead to progress in science. Certainly, a number of such intellectual conflicts were played out in the pages of the *Archives* while he was editor. The disagreements between Roux and Driesch are quite familiar and are well represented in their *Archives* papers. Less well known are controversies that these two passionate scientists were involved in with others; some of these are also discussed below. These often heated exchanges give an idea of the more personal and immediate nature of scientific writing at the turn of the century.

In addition to experimental embryology, several other broad categories can be defined among papers in the *Archives*. These provide some idea of the questions that occupied experimental biologists during the last part of the nineteenth and early twentieth centuries. We will discuss briefly a few of these topics: chemical embryology/developmental physiology; developmental endocrinology; models of life processes; and biomechanics.

Studies on regeneration in a wide range of organisms constitute a significant segment of the papers that appeared while Roux was editor of the *Archives*, but these will not be considered here. However, regeneration studies during this period have recently been discussed with reference to developmental biology by Churchill (1991), Dinsmore (1991) and Maienschein (1991b). The difficulty in drawing a clear distinction between experimental embryology and regeneration is apparent in their discussions.

Among other categories that figured prominently in the *Archives* between 1894 and 1924 that will not be further considered are cytological studies, including models of cell division and mechanisms of mitosis; genetics and hybridization studies; tissue grafting and tissue explantation; descriptions of "naturally occurring developmental abnormalities"; and many general essays and theoretical papers. In the first two decades of the twentieth century, interest in the inheritance of acquired characteristics was renewed by research published in the *Archives* by a young Austrian biologist, Paul Kammerer. Just before Roux's death, papers on "mitogenetic energy" began to appear from the laboratory of Alexander Gurwitsch in Russia. This work, unlike that of Kammerer, excited immediate controversy, and many papers on this subject, both for and against, appeared in the *Archives* in the next few years.

Some of these areas quickly developed into independent disciplines, while others fell from favor, were disproved, or quietly disappeared. Others, most notably perhaps biomechanics, have recently been reinvigorated.

"Truth is born of conflict"

Roux and Driesch

The stage for the Roux-Driesch controversy was set before publication of the *Archives* began. Driesch, 17 years Roux's junior, had read Roux's 1888 paper (Driesch 1951) while a student at Jena, for an informal student seminar group he and his friend Curt Herbst started to discuss the current literature. (Their professor, Ernst Haeckel, professed he did not believe in reading current literature.) Driesch was excited by the experimental approach and decided to repeat Roux's experiments on the much tougher, transparent and rapidly developing eggs of the sea urchin. The next summer, in Trieste, he separated early blastomeres of sea urchin embryos by vigorous shaking, left the isolated blastomeres in petri dishes, and left for the day. According to his posthumous biography (Driesch 1951), he was startled the next morning to find, not the half-larvae he expected, but complete miniature plutei. Sander (1991a–c, 1992a–c, 1993a) has provided excellent discussions of what followed and how the Roux and Driesch conflict developed.

While scientific arguments between Roux and Driesch lasted for years, and were played out in many papers, a flavor of their style can be gained from papers in Vol. 4 of the *Archives*. The starting point for this particular bout was the publication of a paper by Driesch (1896b) on egg organization and its genesis. Roux responded with two papers: "*Corrections...*" (1896b), and a critique of Driesch's (1894) "*Analytical Theory of Organic Development*" (Roux 1896c). He opens the first: "Driesch's representation of my concepts have contained so many errors, that it seems necessary once again to provide some clarification." In the second paper Roux argues, with justification, that Driesch often gives conflicting explanations for the same phenomenon within the same publication, and later picks and chooses what he wishes as his point of view. Driesch (1897a) immediately replied with "a few words of explanation." The tone of the paper is apparent even in a footnote on the first page: "If I deal here briefly with Roux's article of some 55 pages, I do so on the grounds that the fine points cannot possibly be of interest to any third party; for my part this will not happen again. If, in the future, I am silent about such criticisms, it should be evident what this silence means." Undeterred, Roux (1897a) published *his* comments on Driesch's response in the pages immediately following. Nor would Driesch remain silent for long.

But Roux valued Driesch for his intelligence and his commitment to experimental studies. In a review (1902b), he wrote that Driesch's *"Organic Regulation"* was a significant contribution to the field of developmental mechanics. While remarking that Driesch, with two exceptions, had ignored Roux's theoretical discussions, which Roux had (naturally) remedied elsewhere (1902a), he added that Driesch's concepts "lead us toward our goals, although not directly. Science often progresses in a zig-zag fashion, but champions of opposing views are cautioned not to wander too far on either side from the ultimate goal." He argued that the examples Driesch had chosen to bolster vitalism were subject to other interpretations, while he omitted those that could not support him "('as he has always done') and holds his arguments to be irrefutable." Roux then makes a generous confession: "At present, both [vitalists and Kantian mechanists] must gather all possible supports for their views; the decision as to the greater worth of one over the other will be made in the future. *This is a painful renunciation* [italics mine]. But it is scientific to put the major questions exactly at the place where our knowledge stops. Although we cannot agree with the specific viewpoint that he adopts, we also recognize that a scientist like Driesch, stimulated by his theoretical speculations [to devise] ingenious experiments and to present his results with energy and skill, can bring lasting rewards to science." Roux was also meticulous about recording in *"Personalia"* every academic promotion and honor accorded Driesch, although he continued to twit the vitalists, or "psychomorphologists" as he sometimes referred to them (Roux 1907, 1908).

Driesch could be equally generous about Roux. He wrote (1898b, pp. 712–714): "What separates Roux and me in our views and aims is far less than what binds us together, and becomes insignificant in view of the commonality of our goal. I consider myself as one with Roux in the manner in which we must approach our goals, in the methods which we must use. We must study development for its own reasons and comprehension, not just to compare one "developmental history" with another. We must invoke a special aid to investigation, the only one that enables us to vary interacting factors with any certainty: this method is that of *experimentation*, as Roux with all justification has called it, *analytical experimentation*. Further, the formulation of any experiment is enlightening only if it leads to further fruitful experimental approaches." In the *Festschrift* for Roux's seventieth birthday, Driesch (1920) acknowledged himself "..Roux's student in methodology although often his opponent in the interpretation of events." Roux and Driesch obviously enjoyed the cut and thrust of their verbal fencing, and there is ample evidence that it affected neither their personal relationship nor their respect for each other as scientists (Driesch 1898b, 1920, 1951; Roux 1902; Churchill 1991).

Roux and Oscar Hertwig

Roux's most unrelenting critic was Oscar Hertwig. Hertwig's book. *"Current and Disputed Questions in Biology"* (1897), had an appendix, *"Critical Comments concerning the developmental mechanical 'natural laws' of Roux,"* which was longer than the "main" text itself. Much of the text also consisted of attacks on Roux (see Churchill on Hertwig's "broadsides" – 1991, p. 129). Roux (1897b) responded with an almost point-by-point refutation of Hertwig's attack. Driesch also (1897b) came to Roux's support, with a well-argued defense of biological experiments, a method Hertwig had come to consider without merit. The Hertwig polemic also induced the distinguished zoologist, Otto Bütschli, to declare his support for the experimental approach (Bütschli 1897).

Although Hertwig never published in the *Archives*, Roux had ample opportunity to mount the attack in his Archive reviews of Hertwig's many books. One example suffices to give a flavor of Roux's treatment of Hertwig (Roux 1909, p. 446b). He contrasts Hertwig's historical survey, *"Development of Biology in the 19th Century,"* with Radl's similar *"History of Biology II: Developmental theories in the 19th Century"*: "In Radl, we have a born historian while in Hertwig, we have the opposite extreme, a pure 'Party Liner,' who is neither capable of nor willing to give objective consideration to ideas not personally palatable nor to accord them their appropriate value."

Some other opponents of Driesch

While the Driesch-Roux controversy is familiar to most developmental biologists, Driesch's conflicts with other scientists, waged in part in the *Archives*, seem to have attracted little attention. Considered here are three of those controversies.

Mosaic vs regulative development: cell lineage studies

The major conflict centered on the apparently contradictory concepts of mosaic and regulative development. Driesch (1895a), like Chabry (Fischer 1992; Sander and

Fischer 1992) studied the development of ascidian blastomeres isolated at the two and four cell stages. "Isolated" blastomeres in both studies were usually blastomeres that survived within the egg membranes following the death of one or more sister cells. Chabry had relied largely on naturally occurring cleavage abnormalities (Sander and Fischer 1992). Driesch attempted to separate blastomeres by shaking, as he had done with sea urchin blastomeres. He found that even brief shaking usually resulted in the death of one or more of the blastomeres, so he simply selected for study those two- and four-cell embryos in which one, two, or three blastomeres, survived. Chabry concluded from his observations that development in ascidians was mosaic in nature. Driesch, in contrast (p. 405), concluded that from these early isolated blastomeres, "there does not develop a half (or 1/4, 3/4), right or left (or anterior or posterior) embryo, but a whole larva of half size, lacking only certain insignificant organs." In discussing other instances where mosaic development had been claimed, he concluded "[while] the result of Roux is to be regarded as accurate and, with certain exceptions, clearly comprehensible, the [comparable claims by] Chun for ctenophores are highly unlikely and fragmentary."

Driesch and Morgan each went to Naples in 1894, to repeat Chun's experiments on ctenophore eggs. This first meeting led to a life-long friendship (Allen 1978). Because their results were identical, they (1895) published jointly. They interpreted the larvae that developed from blastomeres isolated at the two cell stage as *partial larvae* but not perfect *half larvae* as Chun had claimed. Driesch and Morgan (1895, p. 220) further concluded that there was protoplasmic determination, but that some regulation also occurred in their experimental animals. "We consider these experiments are the first to provide **direct** evidence of protoplasmic determination". While Chun (1895) and Roux (1895b) responded with heat to their paper, E.B. Wilson (1896, pp. 24–25) recognized how important their results were: "The admission [by Driesch and Morgan] *that the degree of cytoplasmic localization in the egg at the beginning of development varies in different forms is of fundamental importance, for upon it turns the entire theory of embryonic development*." [Italics mine.] Driesch and Morgan, and Wilson also, stressed that the likely source of cellular specification in early development was cytoplasmic in origin, and not nuclear as Roux and Weismann held. Wilson (1903) subsequently also studied early ctenophore development. He concluded that cellular specification in ctenophore embryos occurred later than in the highly mosaic ascidian eggs on one hand and earlier than in the regulative eggs of urchins on the other.

Driesch repeatedly attacked the work of American cell-lineage workers in series of essays (1894, 1895b, 1896a, 1897c) and several longer articles (1898b, 1902, 1905, 1906). Wilson (1896, p. 23) commented on Driesch's misinterpretation of his work in the 1894 treatise *"Analytical Theory of Development,"* by providing side-by-side examples of what he had *in fact* said, and what Driesch *claimed* he had said. Wilson graciously conceded "This misrepresentation was of course unintentional but it is unfortunate he did not read more carefully." E.G. Conklin (1906) wrote an outraged and detailed response to Driesch's (1905) criticisms of his ascidian studies, and his introduction provides an excellent summary of the studies and controversies related to ascidian development after Driesch's 1895 paper. Conklin ended his paper (p. 750): "Driesch is convinced [that individual blastomeres of the ascidian egg, up to the four-cell stage, may give rise to entire larvae]. I have presented evidence that they do not, and the only question of importance in this conflict of opinion is as to the relative weight and value of the evidence offered by each of us." (p. 750).

Debates on vitalism

Driesch came into conflict with C.M. Child and Otto zur Strassen (among others) in claims for "proof" of vitalism (cf. Sander 1992c, 1993a) and his interpretation of these experiments in support of that concept.

The conflict with Child centered on his regeneration experiments in *Tubularia* (Child 1907a, b, 1909). Driesch interpreted Child's results as supporting his idea of "harmonious equipotential systems," while Child interpreted them in the light of his evolving concept of gradients (1913) that would culminate in his book, *"Patterns and Problems of Development"* (1941). Child's response to Driesch's criticisms provides an excellent example of how many scientists must have reacted to Driesch's supreme self-confidence (see also Goldschmidt 1956, p. 68; Sander 1992b, c, 1993a, b). Child's refutations culminate in *"Some corrections and criticisms"* (Child 1907b), from which the following quotation (pp. 145–146) is taken: "It is only to be expected that my views appear 'false and really not worth discussion' to Driesch nor do I expect his opinion to be altered by the present writing. The purpose of these comments is to accomplish what Driesch's [1905] review has not, viz. merely to place my views in their proper light before others, to whom such knowledge as Driesch possesses has not yet been granted, and to whom the question as to whether teleological factors exist in regulation still appears open to discussion. I believe we may with complete equanimity of mind leave Driesch to contemplate the universe from the height of his lonely scholastic eminence. We have but one request to make, and that is that when he considers the results of our labors he shall at least acquaint himself fully with those results and not merely a part of them. [Elsewhere] Driesch remarks: 'I cannot resist making the general comment that these highly esteemed American authors would add to the merit of their contributions if they would only carefully read and assimilate the foreign literature.' (Driesch 1905, pp. 232–233). If Driesch would only take this good advice to himself!"

This of course did not end their argument. Driesch (1908) responded that it was "difficult to argue with an opponent [Child] over a point when once again the opponent doesn't understand a harmonious equipotential system" (p. 407).

Driesch chided zur Strassen for simplified interpretations of Driesch's views in his *Ascaris* papers (1896, 1898). Moreover, zur Strassen's "discussion [of development of giant eggs of *Ascaris*] is outspokenly mechanistic, it is even orthodox Darwinistic" (p. 415). Driesch concluded that it seemed to him "as though the *Ascaris* embryo is a harmonious system, whose potency is masked because a certain physical consistency causes irregularity" (p. 417) – surely not one of Driesch's most incisive and convincing "explanations." zur Strassen's response (1908) was more measured than those of many of Driesch's targets. He said that no one now doubted the existence of harmonious equipotential systems, but he did not believe they were in any way contradictory to mechanistic interpretations. zur Strassen concluded: "Today, authors do not read carefully and with an open mind, the major papers of others, ..[nor]..study them until their every nook and cranny is accessible."

Boveri and Driesch

A third series of disputations carried out in the *Archives* involved Boveri and Driesch. These concerned their differing conclusions about the organization of the sea urchin egg (Sander 1993b) and the extent of paternal contributions to the phenotype of hybrid sea urchin embryos (Boveri 1895, 1903, 1904; Driesch 1898a, 1903a, b). Their exchanges are more restrained and civil in tone, but may explain in part Boveri's remark to Spemann that Driesch was looking at his objects "from far above".

Driesch clearly clung tenaciously to his views. A cursory glance at any of the three critical reviews in the *Archives* (Driesch 1895b, 1896a, 1897c) reveals that he, like Roux, was never averse to self-citation when discussing the work of others.

Chemical embryology/developmental physiology

Curt Herbst is given credit for founding "chemical" experimental embryology (Oppenheimer 1991) with his studies of the effects on development of altering the composition of the culture medium. In 1896, he published his first paper (Herbst 1896a) in the *Archives* on this topic (for references to earlier papers, see Note 36, p. 86 in Oppenheimer). In it he described the effects of lithium on gastrulation, a major contribution that is part of the lore of developmental biology. He followed this with a series of papers under the general title *"Inorganic compounds required for the development of sea urchin larvae, their role, and their substitutes"* (1897, 1901, 1904). However, Herbst's significant 1900 paper on the separation of blastomeres and tissues in calcium-free media, although published in the *Archives,* was not included in this series.

Herbst had diverse biological interests and was a prolific contributor to the *Archives*: between 1894 and 1924 he contributed 28 papers, many of them over 50 pages in length. (His longest, published in 1901, was 214 pages.) He investigated heteromorphic regeneration, his first paper on this subject published in 1896 (Herbst 1896b) and his last – during the period under consideration – in 1916 (for references see Note 74, p. 88 in Oppenheimer). Another interest was parental contributions to the phenotype of hybrid sea urchin larvae, described in a third series of papers, *"Studies on Inheritance"* beginning in 1906, the last in the series while Roux was editor, being published in 1914. He (1924) also conducted experiments designed to determine whether rearing amphibians in different colored backgrounds produced alterations in color and pattern, as claimed by Paul Kammerer (see concluding section).

Herbst was not the first investigator to have an interest in the developmental effects of the culture medium (Oppenheimer 1991), nor was he the only investigator in this period to study the subject. Another noted physiological embryologist of the period, Jacques Loeb, contributed more than twenty papers, the majority short notes, to the *Archives* between 1895 and 1924. These papers dealt with a wide range of subjects: artificial parthenogenesis (e.g., 1902, 1914) for which Loeb is best known; studies similar to those of Herbst on the effects of alteration of culture media (1898, 1909, 1911); the physiology of fertilization (1908); and the role of the central nervous system in metamorphosis and regeneration. T.H. Morgan (1899, 1900) studied the effects of salt solutions and "other agents" on the eggs of *Arbacia* and "other animals." The English embryologist, J.W. Jenkinson, who in 1909 published one of the first books in English on experimental embryology, contributed three papers to the *Archives* (1906, 1910, 1911) on the effects of modifications of the medium on development of frog eggs and tadpoles. The most interesting developmental effects were obtained by Stockard (1907) who produced cyclopean *Fundulus* embryos by adding magnesium to seawater. Following Herbst's lead, he studied the effects of lithium chloride on fish development, a paper published in the *Journal of Experimental Zoology* and abstracted in the *Archives* (1906), and of alcohol on mammalian eggs and embryos (Stockard and Craig 1913). While many embryologists immediately adopted the use of calcium-free seawater to separate blastomeres, only one paper in the *Archives,* that of Maas (1906) on calcareous sponges, concerns effects of calcium-free medium on adults and younger postembryonic stages.

Developmental endocrinology

Beginning in 1910, an increasing number of studies on the developmental effects of hormones appeared in the *Archives*. In 1912, Gudernatsch published a seminal paper on the developmental effects of feeding endocrine organs to tadpoles. This paper was followed by many others dealing with the effects of feeding endocrine glands, especially thyroid and thymus, on growth and metamorphosis of amphibian embryos (Romeis 1913–1915), on extirpation of glands (Adler 1914), and extirpation of glands followed by reversal of the effects by feeding appropriate glands (Schulze 1922). Attempts were also made to affect insect development by feeding vertebrate endocrine organs, with little success (Romeis and Dobkiewicz 1920). Ceni (1921) and de Lisi (1921) used partial decerebration to demonstrate "higher thyroid centers" in the brains of birds and mammals, while observing no such effects in turtles.

The role of gonadal hormones in the differentiation of secondary sexual characteristics and behavior also became an active field of investigation during this period. Several approaches were utilized, including study of physical and behavioral traits, and gonadal structure in experimentally created hermaphrodites, intersexes, castrates, castrates provided with implanted gonads of the opposite or both sexes, and homosexual males.

Considerable ingenuity was exhibited by the Austrian scientists, Julius Tandler and Siegfried Grosz, who approached the problem mainly by studying the effects of castration. Their first paper (1909) was a detailed description of the skeleton of a eunuch from Zanzibar who had died of tuberculosis in a Viennese hospital. Their second study (1910) is interesting from both scientific and sociological viewpoints. In the middle of the eighteenth century an extreme religious sect, the Skopsi, arose in Russia. (The name derives from the Russian word for eunuch: the distinguishing characteristic of the self-mortifying sect was self-castration.) Persecuted in Russia, many members of the sect fled to other countries, and Tandler and Grosz studied men belonging to a colony of exiled Skopsi in Bucharest. Although the Skopsi were mistrustful of others, Tandler and Grosz succeeded in persuading five males, ranging in age from 20 to 42, to provide detailed physical histories. Because of variation within the group in the age at which castration had taken place – three were between 5 and 10 years old when castrated, and two, between 16 and 21 – they provided an opportunity to relate the extent to which physical traits and libido were affected when castration occurred before and after puberty. In 1910, Tandler and Keller described physical changes in young female calves castrated to provide work animals. In a third study (1911, 1913), Tandler and Grosz examined seasonal variation in the testes of the mole, *Talpa europa*, which has only one mating season a year. They were able to demonstrate a correlation between mating activity, the developmental state of the Leydig cells, and spermatogenesis.

E. Steinach published several papers on developmental endocrinology in the *Archives*. The first of these (1916) was a study of body structure and behavior in experimentally created "gynandromorphs" – castrated neonatal, juvenile or adult guinea pigs that subsequently received interstitial implants of both ovaries and testes. Such animals, especially those castrated shortly after birth, later developed a mixture of secondary sexual traits. He used the term "puberty glands" for the somatic elements of the gonads, to stress their hormonal functions. In the same volume, he and Holzknecht (1916) reported on enhanced effects of the sex hormones in both sexes when hormone-producing tissues were hypertrophied as a result of experimental intervention. Later papers (1920a–c) dealt with analogous developmental effects of experimentally produced and naturally occurring altered balance of homologous or heterologous sex hormones; histology of gonads of homosexual males; and on rejuvenation through experimental reactivation of aging puberty glands. In this last paper, Steinach included reports of three cases in human males in whom physical rejuvenation was claimed to have occurred following surgery that led to reactivation of interstitial cells. Also in 1920, a paper by Steinach and Kammerer, *"Climate and Masculinity"*, appeared, reporting experimental studies on the effects of elevated temperature on gonadal and scrotal development in rats, and an "anthropological" section, utilizing reports of others on sexual development in various primitive people in the tropics.

Wagner (1922) and his coworkers Lipschütz and Ottow (1922) investigated eunuchoid rabbits and established a relationship between development of eunuchoid characteristics and reduced activity of Leydig cell. Lipschütz earlier had published a study on experimental masculinization (1918a), and two reviews (1918b, c) on the general theory of the role of the puberty glands in production of secondary sexual characteristics. He also published on the relationship between gonads and sex-specific characteristics (1923a) and on developmental effects of unilateral castration (1923b). In work related to Lipschütz' studies on experimental masculinization, Stein and Herrmann (1921) investigated the effects of injection of female hormones on development of secondary sexual characteristics in males.

Kitahara (1923) described in detail the histology of the Leydig cells in man and other mammals, and discussed their physiological function.

Models of life processes

Because the long-term goal of "Kantian Mechanists" was to explain biological functions in terms of the laws of physics and chemistry, model building was a popular exercise. Nor was interest in inanimate systems that "behaved" like living organisms limited to the world of the laboratory scientist. Many such experiments described in the popular scientific press in Germany were simple enough to be carried out by laymen at home with easily available materials (cf. Mann 1948, pp. 13–20).

The foremost modeler of the time was Ludwig Rhumbler, who published many papers in the *Archives*. For an overview of physical and chemical model systems devised during this period, readers are referred to Rhumbler's (1921) comprehensive review. For developmental biologists, a paper that merits study is his long paper (1902) on the course and forces of gastrulation in a wide range of animals, based on a study of slides (some provided by Roux) and of figures and descriptions from the papers of others. One of the few British scientists to publish in the *Archives* during its first three decades, Richard Assheton, also published (1910) a theoretical paper on gastrulation.

Another model builder was R.E. Liesegang (1911, 1912). Speculative physical models of mitosis and cytokinesis, often using comparisons with electromagnetic forces, were also popular, but none of these models has stood the test of time.

Biomechanics

Roux had an early and abiding interest in biomechanics. His earliest publications (annotated in the Collected Works, Vol. I, 1895a), were biomechanical papers, beginning with studies on physical forces responsible for the branching of blood vessels, the topic of the thesis for his medical degree at Jena (p. 1ff). He studied structure as it related to function in the highly differentiated connective tissue in the tail fin of dolphins (p. 458ff.); "self-regulation" of skeletal muscle length in humans (p. 575ff.); and described and speculated on the developmental origins of bony ankylosis of the knee (p. 662ff.). After his move to Halle, this interest was reflected in the research of his colleagues and students (Roux 1920, 1923). Roux in his professorial role taught Human Anatomy as a living subject, most effectively during dissection demonstrations where he and his assistants engaged in dialogues on the relationship between the structure, form and function of various tissues and organs (Roux 1923; Barfurth 1925). First among his Halle colleagues in Roux's esteem and affections was surely Walter Gebhardt, who joined Roux's Anatomy Institut in 1901, initially as an unpaid assistant (Roux 1918). Gebhardt's first paper (1900), based on his thesis supervised by Gustav Born, was on the structure of teeth as reflected by their specific functions. In his major work (1901, 1902, 1905), he correlated gross and fine structure of bones with their functions. Two more theoretical papers dealt with the physical factors that produced specific bone shapes and structure (1903) and a comparison between bone formation and the chemistry of colloids (1911).

The relationship between skeletal structure, form and function was a popular topic, from Tornier's paper (1894–95) on the origin of joints to Weidenreich's (1922) discussion on forces determining skeletal form and the heritability of shape of bones. Shortly after Tornier's paper appeared, Hammar (1896) published an essay on mechanical factors involved in joint formation. The pelvic girdle was used by von Arx (1910) to consider the static principle in the construction of the body. Even small skeletal variations provided opportunity for an examination of morphogenetic forces, as in Morita's (1913) paper on the physical factors that determined the form of vertebral processes. Naturally occurring abnormalities also provided material for study as in Wiesermann's (1908) paper on the mechanical causes of *Chondrodystrophia foetalis*. Dammann (1908) used still another approach, that of comparative studies, to discuss the structure and functional adaptation of tendons.

Experimental studies were also conducted. Fuld (1901) studied changes in the structure of the hind legs of dogs following loss of, or injury to, the forelegs as an example of forces shaping bone structure. Levy (1904), a student of Roux, examined the formative role of tension on tendons. Another student, Schepelmann (1907), investigated form changes induced in the internal organs of geese fed different diets.

Mirroring another early interest of Roux are papers by Thoma (1901), Hess (1903) and Spitzer (1919, 1921) on mechanical forces that affect the structure of the circulatory system. A related paper is that of Spalteholz (1923) on circulatory patterns and organ formation.

The papers of two contributors to the Archives seem quite "modern" biomechanical studies. Petersen (1914, 1924) focused on comparative and general mechanics of animal skeletons. His first paper dealt with the structure and function of the temporomaxillary joint of the cod. The title of the second, *"Concerning the functional structure of the wing bones of the bat, Vespertilio murinus and also the deformation* [Einknicken] *of tubes on bending"* conveys its general theme. Wegner (1915) studied the relationship between the anatomy, statics and mechanics of the spine of the horse. He likened its structure to that of a suspension bridge and used force diagrams to bolster his thesis. He also published Muybridge-like sequential pictures of moving horses to support his arguments.

Competing journals

By 1904, the growing number of experimental zoologists in the United States stimulated the American Zoological Society to establish the *Journal of Experimental Zoology* (JEZ) (Appel 1988; Maienschein 1991c). Only three inconsequential papers by foreigners appeared in the JEZ during its first decade (the period considered here). Many Americans, especially Morgan, his students, and Child had been the most regular early contributors to the *Archives*. It comes as a surprise to discover that the existence of the JEZ, during its first decade, did not have a significant impact on American contributions to the *Archives.* Between 1905 and 1910, Morgan contributed nine papers to the *Archives*, and eleven to the JEZ. After 1910, however, his interests turned entirely to genetics. He and his students published many of their genetics papers in JEZ, but none in the *Archives.* Until her death, Nettie Stevens continued to contribute to the Archives, but her memorable papers, and those of Wilson, on the sex chromosomes of insects, appeared in JEZ. Child, a prolific contributor to many journals, was completely even-handed during this decade, publishing 17 papers in the Archives and 17 in the JEZ. A further 40 American contributors published almost 90 papers in the Archives between 1905 and 1914. For his part, Roux appeared to welcome the JEZ as a sign of the health of the new science: detailed abstracts of the papers in its first four volumes, many of them expanded abstracts by the original contributors, were published in the *Archives*, as were subsequent author abstracts from later volumes. Most of the protozoology papers abstracted by Rhoda Erdmann (1913, 1914) in the *Archives* were published in the JEZ.

Beginning in the second half of the nineteenth century, the number of new journals competing for papers in experimental biology grew rapidly in both Europe and the United States (Gascoigne 1985; Appel 1988). However, the first major new journals dedicated solely to developmental biology – the *Journal of Embryology and Experimental Morphology* (now *Development*) *and Developmental Biology* – did not begin publication until 1953 and 1959, respectively.

Classic papers, major figures

The *Archives* had a considerable impact on the emerging discipline of experimental embryology during its first three decades. It seems appropriate to consider in conclusion some of the most significant contributors and contributions while Roux was editor.

The seminal papers of Roux (1888) and Driesch (1892) appeared before the *Archives* began publication (Sander 1991c, 1992a; abbreviated translations in Willier and Oppenheimer 1974). However, two interesting *Archives* papers by Roux (1894b, 1896a) on the behavior of isolated blastomeres provided the inspiration for Holtfreter's important studies on tissue affinity (1939; translated in Willier and Oppenheimer 1974).

Many papers whose context, if not content, is known to present-day developmental biologists appeared in the *Archives* during its first three decades. Herbst's papers on lithium-induced exogastrulation (1896a) and on separation of cleavage cells in calcium-free seawater (1900), a technique utilized by generations of embryologists, have been referred to earlier. Born's (1896–1897) technique of heteroplastic and xenoplastic transplantation in amphibians (*Rana, Bombinator*) provided an essential method for the important studies of many noted embryologists, including Harrison (1898), O. Mangold (1920, 1924), and Spemann (1921 and later papers) and his students, many of whose papers appeared in the *Archives* in the next decade and a half (Hamburger 1988).

Boveri's most significant papers were published elsewhere, but an early paper (1895) on fertilization and development of enucleated sea urchin eggs, and the fertilization of such eggs with the sperm of related species certainly spawned a whole school of papers. F.R. Lillie's first paper on differentiation without cleavage in *Chaetopterus* appeared in the *Archives* (1902), although his later papers on this topic are more often cited. Finally, the description by Wolff of the phenomenon that has come to be called "wolffian regeneration" of the urodele lens was published in the *Archives* (1895, 1901).

There are also papers that, while not "classics," are landmarks of a sort. Between 1895 and 1914, T.H. Morgan published 27 papers in the *Archives*, two with Driesch, and three short notes with students (his Bryn Mawr students published about 20 papers during the same period), but no one of these stands out as a classic contribution. However, he published an extensive series of informative experiments under the general title *"The relation between the normal and abnormal development of the embryo of the frog"* (1902–1905; Morgan and Torelle 1904). He repeated Roux's experiments, verifying Roux's observations but giving a different interpretation the results. He also performed related experiments, removing upper blastomeres at later cleavage stages, injuring only the top of the egg during early cleavage, etc. In the last paper in the series (X), Morgan discussed the early stages of normal development in light of the results of experimental studies, and how these clarified some of the then current misconceptions concerning normal embryogenesis. His descriptions in this paper of the generation of cell shape changes and their relation to morphogenesis are also of interest.

Wilson's (1903) beautifully illustrated experimental analysis of ctenophore development that helped to clarify the discrepant conclusions of Chun and of Morgan and Driesch, is a fine example of the clarity of his thought, the design and execution of experiments, and the elegance with which he expressed his ideas.

Some early papers of zoologists who were to become better known during the next two decades appeared in the *Archives* shortly before Roux's death. The first of Richard Goldschmidt's papers on geographic variation in the moth *Lymantria* was published in 1924 – it has the distinction of being, at 245 pages, the longest paper published in the *Archives* in its first 30 years. Paul Weiss also published the first of his many papers on transplantation and regeneration of amphibian limbs in the *Archives*, in 1923. A theoretical paper by Gurwitsch (1922) on the concept of embryonic fields was to be taken over and expanded by Weiss later in his career. The distinguished Swedish embryologists J. Runnström and S. Hörstadius also published in the *Archives* during the period under consideration, but not those experimental studies for which they are best known.

A series of papers that became famous in quite another context, not only to the scientific world but to the general public through *"The Case of the Midwife Toad,"* a book by Arthur Koestler (1971), also appeared in the *Archives* during this period. These were by the brilliant young Austrian scientist, Paul Kammerer, who claimed to have achieved the transmission of environmentally acquired traits to offspring. He was subsequently accused of scientific fraud when one of this specimens was found to be "doctored." Unfortunately, a careful examination of his papers and claims, although providing new evidence, is precluded here by lack of space.

The most significant paper that appeared in the *Archives* between 1894 and 1924 is without doubt that of Spemann and Hilde Mangold (1924) on the Organizer. And Spemann must surely be accorded as first among his peers during this period, not only for this many scientific contributions, but also for the students he trained whose papers would subsequently grace the pages of the *Archives*. [For recent discussions of Spemann and his contributions, see Horder and Weindling 1986; Hamburger 1988; Saha 1991.]

Each of Spemann's contributions to the *Archives* between 1901 and 1924 contains illustrations and information that are familiar today, although their source is not always acknowledged. His descriptions of methods and tools for experimentation have been an invaluable resource, primary or secondary, for generations of embryologists. Read in sequence, these papers illustrate Spemann's maturation as an experimental embryologist and the evolution of his experiments and ideas. And by the time all of the papers have been examined, the reader is familiar with the detailed "life histories" of many of the individual embryos that formed the basis of his conclusions.

His first *Archives* papers (1901–1903) were on the effects of constriction of eggs and early embryos of *Triton*. In these papers, he showed that isolated amphibian blastomeres could develop into whole embryos. He found that partial constriction at the two-cell stage, when the cleavage plane was median, produced duplications or twins, depending on the tightness of the ligature. However, if the cleavage plane were frontal, only the dorsal half developed into an embryo, and the ventral into a "belly piece" (Bauchstück) containing all three germ layers but no axial organs. This provided the first evidence of spatial "informational organization" in the amphibian egg. It may come as a surprise that use of the hair-loop constriction technique did not originate with Spemann. Similar methods had been developed by O. Hertwig (1892), Endres (1895), Endres and Walther (1895), and Herlitzka (1895, 1897); all are cited by Spemann.

Spemann's next paper appeared in the *Archives* in 1910, in the *Festschrift* volume for Roux's 60th birthday, and concerns the developmental capacity of inverted otic discs. It is notable for the detailed descriptions of his operative techniques: they are so well and thoroughly described it would be possible today to replicate Spemann's experiments exactly.

Eight years elapsed before Spemann again published a research paper in the *Archives* (1918), a study of the determination of the first primordia in the amphibian embryo. By this time, Spemann had found that newts were preferable to frogs as experimental animals. Not only were their eggs more highly resistant to damage, but different females laid eggs of different color. These color differences, retained beyond neurulation, provided a natural marker to distinguish between host and donor contributions, clearly foreshadowing his decision to use heterospecific grafting. In this paper, he described and detailed additional methods for making the tools he used. Far more significantly, he here concluded that "determination, or its 'precondition' is first present in the dorsal lip and spreads from there" (p. 483). He was cautious about over-interpretation of experimental results: "We must not forget that [these results] reveal only the 'potential' of the developing organism. If they seem of significance in explaining normal development, then possibly we can consider them to constitute cardinal processes in development" (p. 495).

In 1921, Spemann published his first paper on transplantation between *Triton taeniatus* and *T. cristatus* embryos. [In retrospect it is surprising that he had not utilized heteroplastic transplantation earlier, since Born had demonstrated in 1896 the feasibility of heteroplastic and xenoplastic transplantation in amphibian embryos.] In this paper, he concluded, "We can call [the dorsal lip of the blastopore] for brevity's sake, an 'Organizer'; it creates, in the indifferent material in which it lies, or in which it is implanted, an 'Organization Field' of determined direction and capable of expansion. Simple experiments will show upon what this depends" (p. 568). In an appendix added in proof, he reported that his student, Hilde Proschöldt had, *at his suggestion*, transplanted the dorsal lip region from a *T. cristatus* gastrula into the undifferentiated ectoderm of a *T. taeniatus* embryo that subsequently developed two neural plates. Since sections had not been examined, it was not known – but clearly anticipated – which cells contributed the secondary axis. These crucial "simple" experiments, suggested in 1921 and carried out by Hilde Mangold nee Proschöldt, led to their classic paper of 1924 (translated in Willier and Oppenheimer 1974). A detailed discussion by Spemann of his experiments and those of this students and coworkers is found in his book, *"Embryonic Development and Induction"* (1938).

Wilhelm Roux the man

Roux has been discussed as the founder and editor of the *Archives* for its first three decades. I will end with some remarks about Roux the man, especially since much of what has been said has perhaps stressed his less admirable traits. He was a man of considerable intelligence, great imagination and scientific insight, apparently gifted with boundless energy, and completely dedicated to his science (Driesch 1920; Spemann 1920) – to the extent that in his autobiography (1923) he makes no mention of his wife and three children. He had a dry sense of humor, which he displayed only among close friends, such as Barfurth (1920, 1925). He was a generous host, his wife set an excellent table and he had a taste for fine wines (Goldschmidt 1956, pp. 10–11).

It is to be hoped that someone will write a biographical sketch that shows that there was much more to Roux than his well-publicized failings. Until then, the short, affectionate biographical sketches by his longtime friend, Dietrich Barfurth (1910, 1920, 1925), and Roux's responses (1910, 1920) to the *Festschriften* for his 60th and 70th birthdays are recommended to provide a more well-rounded sense of Wilhelm Roux. His autobiography (1923) is often difficult reading, but there are also flashes of his wry wit and nicely ironic sense of humor.

Acknowledgements All translations are by the author; square brackets indicate author's insertions. I must acknowledge the wisdom of the late Professor A.S. Pearse, first chairman of the Zoology Department at Duke, in purchasing in 1926 for $ 600, the first 52 volumes of Roux's Archives. The staff of the BioScience Library at Duke have earned my gratitude for their patience, help and indulgence. Mr. James Lin, an undergraduate assistant, copied the tables of contents of the Archives from 1894 through 1924, provided each author's location and, when possible, the date of submission of manuscripts. His labors contributed greatly to this paper.

References

Note. In view of problems with computerized referencing (see the Editorial on page E1 of Vol. 199), all papers that were published in this journal are listed under its present name. The previous names were *Archiv für Entwicklungsmechanik* until Vol. 43/1918; *Wilhelm Roux' Archiv für Entwicklungsmechanik der Organismen* until Vol. 52/1923 and again from Vol. 105/1925 to Vol. 177/1975; inbetween combined with the *Archiv für mikroskopische Anatomie* from Vol. 52 (re-numbered as Vol. 97) until Vol. 104; *Wilhelm Roux's Archives of Development Biology* until Vol. 195/1985–86; since then *Roux's Archives of Developmental Biology*. When parts of a paper or series of papers were spread over several years, the different years and volumes/pages are listed in sequence but separated by semicolons.

Adler L (1914) Metamorphose studien an Batrachierlarven. I. Exstirpation endokriner Drüsen. Roux's Arch Dev Biol 39:21–45; 40:1–17, 18–32

Allen GE (1978) Thomas Hunt Morgan. The man and his science. Princeton University Press, Princeton

Appel TA (1988) Organizing biology: the American Society of Naturalists and its "Affiliated Societies," 1883–1923. In: Grainger R, Benson KR, Maienschein J (eds) The American development of biology. University of Pennsylvania Press, Philadelphia, pp 87–120

Arx M von (1910) Der Mechanismus des Beckenbodens und das statische Prinzip in Aufbau unseres Körpers. Roux's Arch Dev Biol 29:325–353

Assheton R (1910) The geometrical relation of the nuclei in an invaginating gastrula considered in connection with cell rhythm, and Driesch's conception of entelechy. Roux's Arch Dev Biol 29:46–78

Barfurth D (1910) Wilhelm Roux zum 60. Geburtstage. Roux's Arch Dev Biol 30 (Part 1):vii–xxxvii

Barfurth D (1920) Wilhelm Roux zum siebzigsten Geburtstage. Naturwissenschaften 8:431–459

Barfurth D (1925) Wilhelm Roux: ein Nachruf. Roux's Arch Dev Biol 104:i–xxii

Born G (1896; 1897) Über Verwachsungsversuche mit Amphibienlarven. Roux's Arch Dev Biol 4:349–465, 517–623

Boveri T (1895) Über die Befruchtungs- und Entwickelungsfähigkeit kernloser Seeigeleier und über die Möglichkeit ihrer Bastardirung. Roux's Arch Dev Biol 2:394–443

Boveri T (1903) Über den Einfluss der Samenzelle auf Larvencharaktere der Echiniden. Roux's Arch Dev Biol 16:340–363

Boveri T (1904) Noch ein Wort über Seeigelbastarde. Roux's Arch Dev Biol 17:521–525

Bütschli O (1897) Bemerkungen über die Anwendbarkeit des Experiments in der Entwickelungsmechanik. Roux's Arch Dev Biol 5:591–593

Ceni C (1921) Das Gehirn und die Schilddrüsenfunktion. Experimentelle Untersuchungen. Roux's Arch Dev Biol 47:587–616

Child CM (1907a) An analysis of form regulation in *Tubularia*.I–VI. Roux's Arch Dev Biol 23:398–456; 24:1–28, 285–349

Child CM (1907b) Some corrections and criticisms. Roux's Arch Dev Biol 24:131–146

Child CM (1909) The regulation of mutilated primordia in *Tubularia*. Roux's Arch Dev Biol 27:106–118

Child CM (1913) Studies on the dynamics of morphogenesis and inheritance in experimental reproduction. VI. The nature of axial gradients in Planaria and their relation to antero-posterior dominance, polarity and symmetry. Roux's Arch Dev Biol 37:108–157

Child CM (1941) Patterns and problems of development. University of Chicago Press, Chicago

Chun C (1895) Bemerkungen über den Aufsatz von H. Driesch und T.H. Morgan, "Von der Entwickelung einzelner Ctenophorenblastomeren." Roux's Arch Dev Biol 2:444–447

Churchill FB (1973) Chabry, Roux and the experimental method in nineteenth-century embryology. In: Giere RN, Westfall RS (eds) Foundations of the scientific method: the nineteenth century. Indiana University Press, Bloomington, pp 161–205

Churchill FB (1975) Wilhelm Roux. Dictionary of scientific biography, vol 11. Scribner, New York, pp 570–575

Churchill FB (1991) Regeneration 1885–1901. In: Dinsmore C (ed) A history of regeneration. Cambridge University Press, Cambridge, pp 111–131

Conklin EG (1906) Does half an Ascidian egg give rise to a whole larva? Roux's Arch Dev Biol 21:727–753

Dammann O (1908) Vergleichende Untersuchungen über den Bau und die funktionelle Anpassung der Sehnen. Roux's Arch Dev Biol 26:349–371

Dinsmore CE (1991) Introduction. In: Dinsmore CE (ed) A history of regeneration. Cambridge University Press, Cambridge, pp1–5

Driesch H (1892) Entwicklungsmechanische Studien I. Der Werth der beiden ersten Furchungszellen in der Echinodermenentwicklung. Z Wiss Zool 53:160–178

Driesch H (1894) Analytische Theorie der organischen Entwicklung. Engelmann, Leipzig

Driesch H (1895a) Von der Entwickelung einzelner Ascidienblastomeren. Roux's Arch Dev Biol 1:398–413

Driesch H (1895b) Neuere Beiträge zur exakten Formenkunde in englischer Sprache I. Kritisches Referat. Roux's Arch Dev Biol 1:414–441

Driesch H (1896a) Neuere Beiträge zur exakten Formenkunde in englischer Sprache. II. Roux's Arch Dev Biol 3:317–338

Driesch H (1896b) Betrachtungen über die Organisation des Eies und ihre Genese. Roux's Arch Dev Biol 4:75–124

Driesch H (1897a) Einige Worte der Erläuterung. Roux's Arch Dev Biol 4:731–737

Driesch H (1897b) Über den Werth des biologischen Experiments. Roux's Arch Dev Biol 5:133–142

Driesch H (1897c) Neuere Beiträge zur exakten Morphologie in englischer Sprache. III. Roux's Arch Dev Biol 5:143–167

Driesch H (1898a) Über rein-mütterliche Charaktere an Bastardlarven von Echiniden. Roux's Arch Dev Biol 7:65–102

Driesch H (1898b) Resultate und Probleme der Entwickelungsphysiologie der Tiere. Erg Anat Entwickelungsgeschichte 8:697–846

Driesch H (1902) Über ein neues harmonisch-äquipotentelles System und über solche Systeme überhaupt. Roux's Arch Dev Biol 14:227–246

Driesch H (1903a) Über Seeigelbastarde. Roux's Arch Dev Biol 16:713–722

Driesch H (1903b) Drei Aphorismen zur Entwickelungsphysiologie jüngster Stadien. Roux's Arch Dev Biol 17:41–53

Driesch H (1905) Neue Antworten und neue Fragen der Entwicklungsphysiologie. Erg Anat Entwicklungsgeschichte 11:784–945

Driesch H (1906) Die Entwicklungsphysiologie von 1902–1905. Erg Anat Entwicklungsgeschichte 14:603–807

Driesch H (1908) Über einige neuere "Widerlegungen" der Vitalismus. Roux's Arch Dev Biol 25:407–422

Driesch H (1920) Wilhelm Roux als Theoretiker. Naturwissenschaften 8:446–450

Driesch H (1951) Lebenserinnerungen. Aufzeichnung eines Forschers und Denkers in entscheidender Zeit. Reinhardt, Munich

Driesch H, Morgan TH (1895) Zur Analysis der ersten Entwickelungstadien des Ctenophoreneies. Roux's Arch Dev Biol 2:204–215, 216–224

Endres H (1895) Anstichversuche an Eiern von *Rana fusca*. II. Roux's Arch Dev Biol 2:517–543

Endres H, Walter HE (1895) Anstichversuche an Eiern von *Rana fusca*. I. Roux's Arch Dev Biol 2:38–51

Erdmann R (1913; 1914) Referate über Experimente an Protisten. Roux's Arch Dev Biol 35:781–787; 37:473–478; 40:151–155

Erdmann R (1920) Endomixis and size variation in pure bred lines of *Paramecium aurelia*. Roux's Arch Dev Biol 46:85–148

Erdmann R (1921) Das Verhalten der Herzklappen der Reptilien und Mammalier in der Gewebekultur. Roux's Arch Dev Biol 48:571–620

Fischer JL (1992) The embryological oeuvre of Laurent Chabry. Roux's Arch Dev Biol 201:125–127

Fuld E (1901) Über Veränderung der Hinterbeinknochen von Hunden in Folge Mangels der Vorderbeine. Beiträge zur Frage nach den Ursachen der Knochengestaltung und zur Transformationslehre der Organismen. Roux's Arch Dev Biol 11:1–64

Gasgoigne RM (1985) A historical catalogue of scientific periodicals, with a survey of their development. Garland, New York London

Gebhardt W (1900) Über den funktionellen Bau einiger Zähne. Roux's Arch Dev Biol 10:135–243, 263–360

Gebhardt W (1901; 1905) Über funktionell wichtige Anordnungsweisen der gröberen und feineren Bauelemente des Wirbelthierknochens. Roux's Arch Dev Biol 11:383–498; 12:1–52, 167–223; 20:187–322

Gebhardt W (1903) Auf welche Art der Beanspruchung reagiert der Knochen jeweils mit der Ausbildung einer entsprechenden Architektur? Roux's Arch Dev Biol 16:377–410

Gebhardt W (1911) Knochenbildung und Colloidchemie. Roux's Arch Dev Biol 32:727–734

Giglio-Tos E (1923) Entwicklungsmechanische Studien. II. Die Coeloblastula. Roux's Arch Dev Biol 98:446–495

Goldschmidt R (1924) Untersuchungen zur Genetik der geographischen Variation. I. Roux's Arch Dev Biol 101:92–337

Goldschmidt R (1956) Portraits from memory. Reflections of a zoologist. University of Washington Press, Seattle

Gudernatsch JF (1912) Feeding experiments on tadpoles. I. The influence of specific organs given as food on growth and differentiation. A contribution to the knowledge of internal secretions. Roux's Arch Dev Biol 35:457–483

Gurwitsch A (1922) Über den Begriff des Embryonalen Feldes. Roux's Arch Dev Biol 51:383–415

Hamburger V (1988) The heritage of experimental embryology. Oxford University Press, Oxford

Hammar JA (1896) Durch welche mechanischen Faktoren wird die Chondroidwandlung in den Gelenken hervorgerufen? Roux's Arch Dev Biol 2:559–563

Harrison RG (1898) The growth and regeneration of the tail of the frog larva. Studied with the aid of Born's method of grafting. Roux's Arch Dev Biol 7:430–485

Heffner B (1908) Über experimentell erzeugte Mehrfachbildung des Skeletts bei Echinidenlarven. Roux's Arch Dev Biol 26:1–46

Herbst C (1896a) Experimentelle Untersuchungen über den Einfluß der veränderten chemischen Zusammensetzung des umgebenden Mediums auf die Entwickelung der Thiere. [III.–VI. Theile]. Roux's Arch Dev Biol 2:455–516

Herbst C (1896b) Über die Regeneration von antennenähnlichen Organen an Stelle von Augen. I. Roux's Arch Dev Biol 2:544–558

Herbst C (1897; 1901; 1904) Über die zur Entwickelung der Seeigellarven nothwendigen anorganischen Stoffe, ihre Rolle und ihre Vertretbarkeit. I. Die zur Entwickelung nothwendigen anorganischen Stoffe. II. Die Vertretbarkeit der nothwendigen Stoffe durch andere ähnlicher chemischer Natur. III. Die Rolle der nothwendigen anorganischen Stoffe. Roux's Arch Dev Biol 5:649–794; 11:617–690; 17:306–520

Herbst C (1900) Über das Auseinandergehen von Furchungs- und Gewebezellen in kalkfreiem Medium. Roux's Arch Dev Biol 9:424–463

Herbst C (1906) Vererbungsstudien. I–III. Roux's Arch Dev Biol 21:173–305

Herbst C (1914) Vererbungsstudien. Die grössere Mutterähnlichkeit der Nachkommen aus Rieseneiern. Roux's Arch Dev Biol 39:617–650

Herbst C (1916) Über die Regeneration von antennenähnlichen Organen an Stelle von Augen. VII. Die Anatomie der Gehirnnerven und des Gehirns bei Krebsen mit Antennulis an Stelle von Augen. Roux's Arch Dev Biol 42:407–489

Herbst C (1924) Beiträge zur Entwicklungsphysiologie der Färbung und Zeichnung der Tiere. 2. Die Weiterzucht der Tiere in gelber und schwarzer Umgebung. Roux's Arch Dev 102:130– 167

Herlitzka A (1895) Contributo allo studio della capacita evolutiva dei due primi blastomeri nell' uovo di tritone. Roux's Arch Dev Biol 2:352–369

Herlitzka A (1897) Sulla svilluppo di embrioni completi da blastomeri isolata di uova di tritone (*Molge cristata*). Roux's Arch Dev Biol 4:624–658

Hertwig O (1892) Urmund und *Spina Bifida*. Arch mikroskopische Anat 39:353– 502

Hertwig O (1897) Zeit- u. Streitfragen der Biologie, vol 2: Mechanik und Biologie. Mit einem Anhang: Kritische Bemerkungen zu den 3 'entwicklungmechanischen Naturgesetzen' von Roux. Fischer, Jena

Hess W (1903) Eine mechanische bedingte Gesetzmässigkeit im Bau des Blutgefäßsystems. Roux's Arch Dev Biol 16:632–641

Holtfreter J (1939) Gewebeaffinität, ein Mittel der embryonalen Formbildung. Arch Exp Zellforsch 23:120–169

Horder TJ, Weindling P (1986) Hans Spemann and the Organiser. In: Horder TJ, Witkowski JA, Wylie CC (eds) A history of embryology. Cambridge University Press, Cambridge, pp 183–242

Horder TJ, Witkowski JA, Wylie CC (eds) (1986) A history of embryology. Cambridge University Press, Cambridge

Jenkinson JW (1906) On the effects of certain solutions upon the development of the frog's egg. Roux's Arch Dev Biol 21:367–460

Jenkinson JW (1909) Experimental embryology. Clarendon Press, Oxford

Jenkinson JW (1910) The effect of sodium chloride on the growth and variability of the tadpole of the frog. Roux's Arch Dev Biol 30:349–356

Jenkinson JW (1911) On the effect of certain isotonic solutions on the development of the frog. A correction and extension of previous observations. Roux's Arch Dev Biol 32:699–716

Kitahara Y (1923) Über die Entstehung der Zwischenzellen der Keimdrüsen des Menschen und der Säugetiere und über deren physiologische Bedeutung. Roux's Arch Dev Biol 52/97:550–615

Koestler A (1971) The case of the midwife toad. Random House, New York

Levy O (1904) Über den Einfluss von Zug auf die Bildung von faserigen Bindegeweben. Zugleich ein Beitrag zur Kenntnis der Sehnenvernarbung. Roux's Arch Dev Biol 18:184–247

Liesegang RE (1911) Nachahmung von Lebensvorgängen. I. Stoffverkehr, bestimmt gerichtetes Wachstum. II. Zur Entwicklungsmechanik des Epithels. III. Formkatalysatoren. Roux's Arch Dev Biol 32:636–650, 651–661; 33:328–338

Liesegang RE (1912) Protoplasmastrukturen und deren Dynamik. Roux's Arch Dev Biol 34:452–460

Lillie FR (1902) Differentiation without cleavage in the egg of the annelid *Chaetopterus pergamentaceus*. Roux's Arch Dev Biol 14:477–499

Lipschütz A (1918a) Umwandlung der Clitoris in ein penisartiges Organ bei experimenteller Maskulierung. Roux's Arch Dev Biol 44:196–206

Lipschütz A (1918b) Prinzipielles zur Lehre von der Pubertätsdrüse. Roux's Arch Dev Biol 44:207–212

Lipschütz A (1918c) Die Gestaltung der Geschlechtsmerkmale durch die Pubertätsdrüsen. Roux's Arch Dev Biol 44:396–440

Lipschütz A (1923a) Zur Frage der geschlechtsspezifischen Beeinflussung der Gonade durch eine heterosexuelle Geschlechtsdrüse. Roux's Arch Dev Biol 52/97:384–385

Lipschütz A (1923b) Beobachtungen zur Frage einseitiger Kastrationserscheinungen. Roux's Arch Dev Biol 52/97:395–399

Lipschütz A, Ottow B, Wagner K (1922) Über Eunuchoidismus beim Kaninchen, bedingt durch Unterentwicklung des Hodens. Roux's Arch Dev Biol 51:66–78

Lisi L de (1921) Über die Funktion der Hoden und des Eierstockes der enthirnten Schildkröten. Roux's Arch Dev Biol 47:617–626

Loeb J (1898) Über den Einfluss von Alkalien und Säuren auf die embryonale Entwickelung und das Wachsthum. Roux's Arch Dev Biol 7:631–641

Loeb J (1902) Über Methoden und Fehlerquellen der Versuche über künstliche Parthenogenese. Roux's Arch Dev Biol 13:481–486

Loeb J (1908) Über die osmotischen Eigenschaften und die Entstehung der Befruchtungsmembran bei Seeigelei. Roux's Arch Dev Biol 26:82–88

Loeb J (1909) Über die chemischen Bedingungen für Entstehung eineiiger Zwillinge beim Seeigel. Roux's Arch Dev Biol 27:119–140

Loeb J (1911) Können die Eier von *Fundulus* und die jungen Fische im destillierten Wasser leben? Roux's Arch Dev Biol 31:654–675

Loeb J (1914) Weitere Beiträge zur Theorie der künstlichen Parthenogenese. Roux's Arch Dev Biol 38:409–417

Maas O (1906) Über die Einwirkung karbonatfreier und kalkfreier Salzlösung auf erwachsene Kalkschwämme und Entwicklungsstadien derselben. Roux's Arch Dev Biol 22:581–599

Maienschein J (ed) 1986 Defining biology. Lectures from the 1890s. Harvard University Press, Cambridge

Maienschein J (1991a) The Origins of *Entwicklungsmechanik*. In: Browder L (ed) Developmental biology, vol 7: Gilbert SF (ed) A conceptual history of modern embryology. Plenum, New York, pp 43–61

Maienschein J (1991b) T.H. Morgan's regeneration, epigenesis and (w)holism. In: Dinsmore CE (ed) A history of regeneration. Cambridge University Press, Cambridge, pp 133–149

Maienschein J (1991c) Transforming traditions in American biology, 1880–1915. The Johns Hopkins University Press, Baltimore

Mangold O (1920) Fragen der Regulation und Determination an umgeordneten Furchungsstadien und verschmolzenen Keimen von *Triton*. Roux's Arch Dev Biol 47:249–301

Mangold O (1924) Transplantationsversuche zur Frage der Spezifität und der Bildung der Keimblätter. Roux's Arch Dev Biol 100:198–301

Mann T (1948) Doctor Faustus. Vintage. Random House, New York

Mayr E (1986) Foreword. In: Maienschein J (ed) Defining biology. Lectures from the 1890s. Harvard University Press, Cambridge, pp vii-x

Morgan TH (1899) The action of salt-solutions on the unfertilized eggs of *Arbacia* and other animals. Roux's Arch Dev Biol 8:448–539

Morgan TH (1900) Further studies on the action of salt-solutions and of other agents on the eggs of *Arbacia*. Roux's Arch Dev Biol 10:489–524

Morgan TH (1902–1905) The relation between normal and abnormal development of the embryo of the frog. I. As determined by injury to the yolk-portion of the egg. II. As determined by the effects of lithium chloride in solution. III. As determined by some abnormal forms of development. V. As determined by the removal of the upper blastomeres of the frog's egg. VI. As determined by incomplete injury to one of the first two blastomeres. VII. As determined by injury to the top of the egg in the two- and four-cell stages. VIII. As determined by injuries caused by a low temperature. IX. As determined by insufficient aeration. X. A re-examination of the early stages of normal development from the point of view of the results of abnormal development. Roux's Arch Dev Biol 15:238–313; 16:691–712; 18:507–534; 19:58–78, 318–347, 566–569, 570–580, 581–587, 588–614

Morgan TH, Torelle E (1904) The relation between normal and abnormal development. IV As determined by Roux's experiment of injuring the first formed blastomeres of the frog's egg. Roux's Arch Dev Biol 18:535–554

Morita S (1913) Über die Faktoren, welche die Richtung und Gestalt der Wirbeldornen bestimmen. Roux's Arch Dev Biol 37:159–182

Oppenheimer JM (1991) Curt Herbst's contributions to the concept of embryonic induction. In: Browder L (ed) Developmental biology: a comprehensive synthesis, vol 7: Gilbert SF (ed) A conceptual history of modern embryology. Plenum, New York, pp 63–89

Petersen H (1914) Studien zur vergleichenden und allgemeinen Mechanik des Tierkörpers. I. Die Kiefergelenke des Kabeljau, *Gadus morrhua*. Roux's Arch Dev Biol 39:51–111

Petersen H (1924) Über den funktionellen Bau der Flügelknochen der Fledermaus (*Vespertilio murinus*) und über das Einknicken von Röhren bei Biegung. Roux's Arch Dev Biol 102:406–425

Rhumbler L (1902) Zur Mechanik des Gastrulationsvorganges insbesondere der Invagination. Eine entwickelungsmechanische Studie. Roux's Arch Dev Biol 14:401–476

Rhumbler L (1921) Methodik der Nachahmung von Lebensvorgängen durch physikalische Konstellationen. Aberhaldens Handbuch der biologischen Arbeitsmethoden. Lief 33

Ridley M (1986) Embryology and classical zoology in Great Britain. In: Horder TJ, Witkowski JA, Wylie CC (eds) A history of embryology, Cambridge University Press, pp 37–65

Romeis B (1913, 1914, 1915) Der Einfluss verschiedenartiger Ernährung auf die Regeneration bei Kaulquappen (*Rana esculenta*). I, II. Der Einfluss von Thyreoidea- und Thymusfütterung auf das Wachstum, die Entwicklung und die Regeneration von Anurenlarven. Roux's Arch Dev Biol 37:183–216; 40:571–652; 41:57–119

Romeis B, Dobkiewicz L von (1920) Experimentelle Untersuchungen über die Wirkung von Wirbeltierhormonen auf Wirbellose. I. Der Einfluss der Schilddrüsenfütterung auf Entwicklung und Wachstum der Schmeissfliege (*Calliphora vomitoria*). Roux's Arch Dev Biol 47:119–130

Rossiter M (1982) Women scientists in America. Struggles and strategies to 1940. The Johns Hopkins University Press, Baltimore

Roux W (1888) Beiträge zur Entwickelungsmechanik des Embryos. Virchows Arch 114:113–153

Roux W (1894a) Einleitung. Roux's Arch Dev Biol 1:1–42

Roux W (1894b) Über den "Cytotropismus" der Furchungszellen des Grasfrosches (*Rana fusca*). Roux's Arch Dev Biol 1:43–68, 161–202

Roux W (1895a) Gesammelte Abhandlung über Entwickelungsmechanik (2 vols). Engelmann, Leipzig

Roux W (1895b) Über die Bedeutung der neuen Versuche an gefurchten und ungefurchten Ctenophoreneiern. Roux's Arch Dev Biol 2:448–454

Roux W (1896a) Über die Selbstordnung (Cytotaxis) sich "berührender" Furchungszellen des Froscheies durch Zellenzusammenfügung, Zellentrennung und Zellengleiten. Roux's Arch Dev Biol 3:381–468

Roux W (1896b) Berichtigungen zu H. Driesch's Aufsatz "Betrachtungen über die Organisation des Eies". Roux's Arch Dev Biol 4:341–343

Roux W (1896c) Zu H. Driesch's "Analytischer Theorie der organischen Entwickelung". Roux's Arch Dev Biol 4:466–501

Roux W (1897a) Bemerkungen zu der vorstehenden "Erläuterung" H. Driesch's. Roux's Arch Dev Biol 4:738–739

Roux W (1897b) Für unser Programm und seine Verwirklichung. Roux's Arch Dev Biol 5:1–80; 219–342

Roux W (1902a) Über die Selbstregulation der Lebewesen. Roux's Arch Dev Biol 13:610–650

Roux W (1902b) Referate. Hans Driesch, Die organischen Regulationen; Curt Herbst, Formative Reize in der thierischen Ontogenese. Roux's Arch Dev Biol 13:651–661

Roux W (1905) Die Entwicklungsmechanik. Ein neuer Zweig der biologischen Wissenschaft. In: Vorträge und Aufsätze über Entwickelungsmechanik der Organismen 1:1–283

Roux W (1907) Besprechungen. Roux's Arch Dev Biol 24:686–695

Roux W (1908) Weitere Bemerkungen über Psychomorphologie und Entwicklungsmechanik. Roux's Arch Dev Biol 25:720–725

Roux W (1909) Besprechung. Roux's Arch Dev Biol 28:446a–446i

Roux W (1910) Worte des Dankes. Roux's Arch Dev Biol 31:i–viii

Roux W (1912) Personalia. [Stevens Obituary.] Roux's Arch Dev Biol 35:586

Roux W (1918) Nekrolog Walter Gebhardt. Roux's Arch Dev Biol 44:652–656

Roux W (1920) Dank. Roux's Arch Dev Biol 47:i-xi

Roux W (1921) Personalia. Roux's Arch Dev Biol 47:668

Roux W (1923) Wilhelm Roux in Halle/Saale. In: Grote RL (ed) Die Medizin der Gegenwart in Selbstdarstellung. I. Meiner, Leipzig, pp 141–206

Saha M (1991) Spemann seen through a lens. In: Browder L (ed) Developmental biology, vol 7: Gilbert SF (ed) A conceptual history of modern embryology. Plenum Press, New York, pp 91–108

Sander K (1991a) Wilhelm Roux and his programme for developmental biology. Roux's Arch Dev Biol 200:1–3

Sander K (1991b) "Mosaic work" and "assimilating effects" in embryogenesis. Wilhelm Roux's conclusions after disabling frog blastomeres. Roux's Arch Dev Biol 200:237–239

Sander K (1991c) Wilhelm Roux and the rest: developmental theories 1885–1895. Roux's Arch 200:293–299

Sander K (1992a) Shaking a concept: Hans Driesch and the varied fates of sea urchin blastomeres. Roux's Arch Dev Biol 201:265–267

Sander K (1992b) Hans Driesch the critical mechanist: "Analytische Theorie der organischen Entwicklung." Roux's Arch Dev Biol 201:331–333

Sander K (1992c) Hans Driesch's "philosophy really ab ovo" or why to be a vitalist. Roux's Arch Dev Biol 202:1–3

Sander K (1993a) Entelechy and the ontogenetic machine – work and views of Hans Driesch from 1895 to 1910. Roux's Arch Dev Biol 202:67–69

Sander K (1993b) Theodor Boveri on cytoplasmic organization in the sea urchin egg. Roux's Arch Dev Biol 202:129–131

Sander K, Fischer JL (1992) How to dart ascidian blastomeres: the embryological microtools of Laurent Chabry (1855–1893). Roux's Arch Dev Biol 201:265–267

Schepelmann E (1907) Über die gestaltende Wirkung verschiedener Ernährung auf die Organe der Gans, insbesondere über die funktionelle Anpassung an die Nahrung. Roux's Arch Dev Biol 23:183–226

Schulze W (1922) Weitere Untersuchungen über die Wirkung inkretorischer Drüsensubstanzen der Morphogenie. II. Neotenie und gesteigertes Wachstum nach Thyroidektomie bei Larven von *Rana fusca.* Wiederbeginn der Fortentwicklung durch Verfütterung von Rinderschilddrüse. Roux's Arch Dev Biol 52/97:232–260

Spalteholz W (1923) Gefässbaum und Organbildung. Roux's Arch Dev Biol 52/97:480–531

Spemann H (1901–1903) Entwickelungsphysiologische Studien am Triton-Ei. I–III. Roux's Arch Dev Biol 12:224–264; 15:448–534; 16:551–631

Spemann H (1910) Die Entwicklung des invertierten Hörgrübchens zum Labyrinth. Ein kritischer Beitrag zur Strukturlehre der Organanlagen. Roux's Arch Dev Biol 30 (Part 2):437–458

Spemann H (1918) Über die Determination der ersten Organanlagen des Amphibienembryo. I–VI. Roux's Arch Dev Biol 43:448–555

Spemann H (1920) Wilhelm Roux als Experimentator. Naturwissenschaften 8:443–446

Spemann H (1921) Die Erzeugung tierischer Chimären durch heteroplastische embryonale Transplantation zwischen *Triton cristatus* and *taeniatus*. Roux's Arch Dev Biol 48:533–570

Spemann H (1938) Embryonic development and induction. Yale University Press, New Haven

Spemann H, Mangold H (1924) Über Induktion von Embryonalanlagen durch Implantation artfremder Organisatoren. Roux's Arch Dev Biol 100:599–638

Spitzer A (1919; 1921) Über die Ursachen und den Mechanismus der Zweiteilung des Wirbeltierherzen. I, II. Roux's Arch Dev Biol 45:686–725; 47:511–570

Stein M, Herrmann E (1921) Über künstliche Entwicklungshemmung männlicher sekundärer Geschlechtsmerkmale. Roux's Arch Dev Biol 48:447–488

Steinach E (1916) Pubertätsdrüsen und Zwitterbildung. Roux's Arch Dev Biol 42:307–332

Steinach E (1920a) Künstliche und natürliche Zwitterdrüsen und ihre analogen Wirkungen. Roux's Arch Dev Biol 46:12–28

Steinach E (1920b) Histologische Beschaffenheit der Keimdrüse bei homosexuellen Männern. Roux's Arch Dev Biol 46:29–37

Steinach E (1920c) Verjüngung durch experimentelle Neubelebung der alternden Pubertätsdrüse. Roux's Arch Dev Biol 46:553–618

Steinach E, Holzknecht G (1916) Erhöhte Wirkung der inneren Sekretion bei Hypertrophie der Pubertätsdrüsen. Roux's Arch Dev Biol 42:490–507

Steinach E, Kammerer P (1920) Klima und Mannbarkeit. Roux's Arch Dev Biol 46:391–458

Stockard CR (1906) The development of *Fundulus heteroclitus* in solutions of lithium chlorid. Roux's Arch Dev Biol 21:359

Stockard CR (1907) The artificial production of a single median cyclopean eye in the fish embryo by means of sea water solutions of magnesium chlorid. Roux's Arch Dev Biol 23:249–258

Stockard CR, Craig DM (1913) An experimental study of the influence of alcohol on the germ cells and the developing embryos of mammals. Roux's Arch Dev Biol 35:569–584

Strassen O zur (1896) Embryonalentwickelung der *Ascaris megalocephala.* Roux's Arch Dev Biol 3:27–105, 133–190

Strassen O zur (1898) Über die Riesenbildung bei *Ascaris*-Eiern. Roux's Arch Dev Biol 7:642–676

Strassen O zur (1908) Zur Widerlegung des Vitalismus. Roux's Arch Dev Biol 26:153–177

Tandler J, Grosz S (1909; 1910) Über den Einfluss der Kastration auf den Organismus. I. Beschreibung eines Eunuchenskeletes. II. Die Skopzen. Roux's Arch Dev Biol 27:35–61; 30 (Part 2):236–253

Tandler J, Grosz S (1911; 1913) Über den Saisondimorphismus der Maulwurfhoden. Roux's Arch Dev Biol 33:297–302; 35:132–134

Tandler J, Keller K (1910) Über den Einfluss der Kastration auf den Organismus. IV. Die Körperform der weiblichen Frühkastraten des Rindes. Roux's Arch Dev Biol 31:289–306

Thoma R (1901) Über den Verzweigungsmodus der Arterien. Roux's Arch Dev Biol 12:352–413

Tornier G (1894; 1895) Das Entstehen der Gelenkformen. Roux's Arch Dev Biol 1:124–158; 307–346

Wagner K (1922) Über die Zwischenzellen und das spermatogene Gewebe in einem Fall von Eunuchoidismus beim Kaninchen. Roux's Arch Dev Biol 51:416–435

Wegner F (1915) Beitrag zur Anatomie, Statik und Mechanik der Wirbelsäule des Pferdes mit besonderer Berücksichtigung der Zwischenwirbelscheiben. Roux's Arch Dev Biol 41:323–369, 371–429

Weidenreich F (1922) Über formbestimmende Ursachen am Skelett und die Erblichkeit der Knochenform. Roux's Arch Dev Biol 51:436–481

Weiss, P (1923) Die Transplantation von entwickelten Extremitäten bei Amphibien. I, II. Arch f. mikroskop. Anat. u. Entwicklungsmech 99:150–167; 168–186

Wheeler WM (1894) Translation of W Roux: The problems, methods and scope of developmental mechanics. Archiv für Entwickelungsmechanik: 1. In: Maienschein J (ed) Defining biology. Lectures from the 1890s. Harvard University Press, Cambridge, pp 105–148

Wiesermann J (1908) Über *Chondrodystrophia foetalis* mit besonderer Berücksichtigung ihrer Entstehung durch mechanische Ursachen. Roux's Arch Dev Biol 26:47–81

Willier BH, Oppenheimer JM (eds) (1974) Foundations of experimental embryology, 2nd edn. Hafner Press, New York

Wilson EB (1896) On cleavage and mosaic work. Appendix to Crampton: Experimental studies on Gasteropod development. Roux's Arch Dev Biol 3:19–26

Wilson EB (1903) Experiments on cleavage and localization in the Nemertine egg. Roux's Arch Dev Biol 16:411–460

Wolff G (1895; 1901) Entwickelungsphysiologische Studien. I. Die Regeneration der Urodelenlinse. II. Weitere Mittheilungen zur Regeneration der Urodelenlinse. Roux's Arch Dev Biol 1:380–390; 12:307–351

MIX
Papier aus verantwortungsvollen Quellen
Paper from responsible sources
FSC® C105338

If you have any concerns about our products,
you can contact us on
ProductSafety@springernature.com

In case Publisher is established outside the EU,
the EU authorized representative is:
Springer Nature Customer Service Center GmbH
Europaplatz 3, 69115 Heidelberg, Germany

Printed by Libri Plureos GmbH
in Hamburg, Germany